The Ageing of Materials and Structures

D1807328

Klaas van Breugel · Dessi Koleva
Ton van Beek

Editors

The Ageing of Materials and Structures

Towards Scientific Solutions
for the Ageing of Our Assets

 Springer

Editors
Klaas van Breugel
Faculty of Civil Engineering
   and Geosciences
Delft University of Technology
Delft
The Netherlands

Ton van Beek
Faculty of Civil Engineering
   and Geosciences
Delft University of Technology
Delft
The Netherlands

Dessi Koleva
Faculty of Civil Engineering
   and Geosciences
Delft University of Technology
Delft
The Netherlands

ISBN 978-3-319-88900-9   ISBN 978-3-319-70194-3   (eBook)
https://doi.org/10.1007/978-3-319-70194-3

Printed on acid-free paper

This Springer imprint is published by Springer Nature
The registered company is Springer International Publishing AG
The registered company address is: Gewerbestrasse 11, 6330 Cham, Switzerland

Preface

Ageing is an inherent feature of nature and of all man-made products. Small consumer goods, electronic devices, power plants, electricity grids, infrastructural works and art treasures—all are subjected to ageing. Although the consequences of ageing can be observed everywhere around us and many strategies have been developed to mitigate these consequences, no comprehensive scientific perception of ageing exists today. Moreover, the omnipresence of ageing and the often very slow progress of ageing phenomena are two reasons why we have got used to the obviously unavoidable presence of ageing. Ageing is a matter of fact, and we have to adapt to it. This attitude, however, seems to be changing lately. More and more, ageing is considered a huge financial burden for modern industrialised societies. Ageing of our assets, assets in the widest sense of the word, is a trillion-dollar issue and a burden for the environment as well. But maybe ageing phenomena also offer new and unexpected possibilities! Ageing not only for worse, but also for better. Whether ageing occurs for either better or worse, in both cases we want to be in full control of ageing phenomena. That's where science and research come in and why the initiative was taken to organise an international conference on ageing of materials and structures, AMS'14.

This book is a collection of papers presented during the 1st International Conference on Ageing of Materials and Structures, AMS'14, held in Delft, the Netherlands. Many lessons were learned from all the contributing authors, from the inspiring and lively discussions, and from the fruitful interactions. This book collects only part of the gained knowledge, emphasising central aspects of ageing of materials and structures.

The essential current state of the art on the subject of ageing is captured in the chapters, reflecting various angles and viewpoints. The chapters present the research focus of the authors on ageing phenomena as embodied in a certain field of interest, but encompassing a wide range of materials and structures relevant to a wide range of applications. This book collects current knowledge on ageing within various scientific fields, specifically addressing the interrelation of scientific background and reflection in practical cases. It links fundamentals to valid real-life perspectives.

The AMS'14 conference was the first international event to reflect the importance of ageing of materials and structures worldwide. The conference targeted rising awareness and enhanced recognition of the subject of ageing. This first international conference is considered a path-finding event, where we want to get ageing phenomena in sharp focus and where we articulate the essence and complexity of ageing of materials and structures.

The initiative for this conference was taken by the Ageing Centre for Materials, Structures and Systems, a spin-off of the Delft Centre for Materials of the TU Delft. The AMS'14 was supported by RILEM (the International Union of Laboratories and Experts in Construction Materials, Systems and Structures). The organisers would like to thank all sponsors of AMS'14 and the respected delegates from all over the world for bringing the conference and this book to a successful conclusion.

Delft, The Netherlands   Prof. Dr. ir. Klaas van Breugel
December 2016

The original version of the book was revised: Author name has been corrected. The erratum to the book is available at https://doi.org/10.1007/978-3-319-70194-3_18

Contents

Part I
Introduction

Urgency and Challenges of Ageing in Science and Engineering

Klaas van Breugel

Abstract Ageing is an inherent feature of nature and, as such, one of oldest performance phenomena observed in the real world. Yet is seems to be a rather new topic in science and engineering. The main reason for this is the growing awareness that ageing of our infrastructure assets threatens the reliability and proper functioning of industrialised societies and is, in the end, a financial burden on the society. In this contribution, the urgency and challenges of ageing phenomena are addressed, including cost aspects and research needs. Emphasis will be on ageing of infrastructure and industrial plants and on the justification of of investments into research on ageing of materials, structures and systems.

Keywords Ageing · Costs · Infrastructure · Risk · Sustainability

1 Introduction

Ageing is everywhere. Huge mountains seem to keep their shape forever. But, at a closer look, we see that the surface of rocks gradually changes. Snow, rain, frost, light, wear, wind and sunshine are sufficiently powerful to crumble even the strongest rock. Mountains age! Tectonic action may fracture mountains, causing rigorous changes in the state of stress in the newly formed parts of the mountain. The fracture surfaces become exposed to climatic conditions and another cycle of ageing starts.

Like rocks, also man-made infrastructures are exposed to climate conditions. While being exposed to environmental loads, structures have to carry deadweight and live loads in a safe way. Even the strongest structures exhibit a decay of quality and function with elapsed time.

K. van Breugel (✉)
Civil Engineering and Geosciences, Delft University of Technology,
Delft, The Netherlands
e-mail: K.vanBreugel@tudelft.nl

© Springer International Publishing AG 2018
K. van Breugel et al. (eds.), *The Ageing of Materials and Structures*,
https://doi.org/10.1007/978-3-319-70194-3_1

Roads and railways need continuous maintenance. If planned correctly, the trouble maintenance will cause can be kept to a minimum. If too late, maintenance and repair will cause time- and money-consuming traffic jams, delays or even accidents.

A proper functioning road network is vital for our economy. Roads are used by cars. The lifetime of cars has dramatically increased in past decades, but still a 20-year-old car is an exception rather than a rule. Electronic systems and sensors may signal a decay of functions, but cannot stop the car's ageing. Moreover, the electronic systems and sensors are subject to ageing themselves, as well.

Cars need fuel, either fossil fuel or electricity. Various fuel types are produced in huge chemical plants, and electricity for hybrid cars is produced in impressive power plants. Fuel and electricity are transported to distribution points. For transport of fuel and electricity, car-carriers, pipelines and electricity grids are used. All these mobile and fixed assets are ageing. Even those with the highest quality sooner or later exhibit ageing and have to be replaced. If not replaced in time, catastrophic accidents may happen.

Power plants for generating electricity, and energy-transport grids have to function reliably 24 h per day, the whole year round. Failing components constitute a risk for life and limb, as well as costly process interruptions. The indirect costs of failures are estimated to be five to ten times the direct costs. Pro-active replacement of vital components of systems and structures is considered a safe strategy to prevent catastrophic failures. But do we really know how close we were to a catastrophic failure at the moment these components were replaced? Were we really at risk or did we spoil a lot of still perfectly operating components without improving safety substantially?

Ageing is everywhere and unavoidable. But if the presence of ageing is unavoidable, a next question is whether the consequences of ageing are unavoidable as well, and whether it is possible to intervene in the rate of ageing. For an answer to these questions, we have to know what ageing really is. Before the ageing issue is discussed in more depth, we will first have a closer look at three different areas where ageing is a hot topic today, i.e. infrastructure, chemical plants and power plants.

2 Ageing and Society

2.1 Ageing Infrastructure

Global infrastructure stock—According to Long [1], the infrastructure of modern industrialised countries makes up over 50% of the nation's national wealth. If this 50% rule would hold for the worldwide infrastructure, a global national wealth in 2015 of US$252 trillion would imply a value of the global infrastructure stock of US$125 trillion. According to Dobbs et al. [2], the value of the infrastructure stock is about 70% of the global gross domestic product (GDP). For a global GDP of US $76 trillion in 2013, this comprises US$53 trillion. In the latter amount, the value of houses is not included, which explains why this value is much lower than the aforementioned value of US$125 trillion.

Our infrastructure consists of roads and railway systems, water works, airports, power stations and telecommunications. This infrastructure is vital for the mobility of people and for a country's economy. Moreover, economic growth is inconceivable without the growth of a country's infrastructure. To catch up with global economic growth, Dobbs et al. [2] estimates a required investment of US$57 trillion between 2013 and 2030. This means an annual investment of US$3.2 trillion, which is about 4.2% of the global GDP (no correction for inflation). The investment of US $57 trillion is needed for roads and railways, ports, airports, power stations, water works and telecommunication. Table 1 gives the breakdown of investments over these categories. These figures are (in part) based on the extrapolation from data provided by 84 countries on expected future investments for infrastructure works. These 84 countries are responsible for 90% of the global GDP. These figures are, therefore, considered today's best possible basis for estimating the extra investments needed for our infrastructure in the indicated period.

Dutch infrastructure stock and fixed assets—In 2012, the GDP of The Netherlands was US$615 billion (Note: adopted exchange rate dollars/euro = 1.11). If the aforementioned 70% rule also applies to the Dutch situation, the value of its infrastructure stock would be US$430 billion. This figure matches well with the official value of the Dutch infrastructure of US$346 billion provided by the Dutch Central Agency for Statistics (CBS) in the year 2009 (in [3]).

The infrastructure makes up only a part of the total fixed assets. The total national wealth of The Netherlands in 2009 was US$4.22 trillion. Almost half of it consists of fixed assets, i.e. infrastructure, houses, industrial buildings and durable capital goods, in total, US$2.025 trillion. From the data in Table 2 it can be inferred

Table 1 Estimated needs for global infrastructure in various categories in the period 2013–2030 to catch up with expected economic growth [2]

Category	Source	Required investment [× US$1,000,000,000,000]
Roads	OECD[a]	16.6
Rail	OECD	4.5
Ports	OECD	0.7
Airports	OECD	2.0
Power	IEA[b]	12.2
Water	GWI[c]	11.7
Telecommunications	OECD	9.5
Total		57.3

[a]Organisation for Economic Co-operation and Development
[b]International Energy Agency
[c]Global Water Intelligence

Table 2 Value of fixed capital goods of The Netherlands, 2009. Total national wealth US$ 4.22 trillion [3]

Fixed capital goods	Value [× US$100,000,000]	Percentage of national wealth (%)
Civil infrastructure	346	8
Houses	1082	25
Industrial buildings	424	10
Permanent capital goods	173	4
Total	2025	47

that Long's statement that the infrastructure makes up about 50% of a nation's national wealth holds quite good for the Dutch situation.

2.2 Ageing of Chemical Plants

Many chemical plants, particularly the larger ones, have attained their present size through a process of gradual expansion. Subsequent parts were built according to various different codes, with different materials and designed while considering different technologies. This process of gradual growth has resulted in a high heterogeneity of large plants. Whereas the typical life cycle of a plant has been estimated at 25 years, the 'effective' age of large plants varies substantially with respect to both its real age and its functional age. Because of this heterogeneity, it is not easy to inspect and judge these plants with respect to their state of ageing. This holds true for both onshore and offshore plants.

The HSE Research Report RR 823 [4] "Plant Ageing Study" gives an overview of ageing phenomena in chemical plants. It covers definitions of ageing and, more importantly, reveals the impact of ageing on plant safety. Based on three principal databases of incidents reports, i.e. RIDDOR, MARS and MHIDAS, the significance of ageing was determined. The MARS study showed that approximately 60% of incidents were related to technical integrity and, of those, 50% have ageing as a contributory factor. From that finding, it was concluded that ageing is a significant issue. The percentage of incidents with ageing as a contributing factor was considered likely to increase with time as assets age. Ageing may affect the installations, piping and containments, as well as the electrical control and instrumentation (EC&I) equipment. Figure 1 shows the results of the MARS incidents with ageing as the cause of failure.

In the UK, 173 loss-of-containment incidents in the period 1996–2008 have been attributed to ageing. This is 5.5% of all loss-of- containment events. Across Europe, 96 incidents occurred due to ageing (MARS database, see [4]). Those incidents represented 28% of all reported 'major accident' loss-of-containment events. In those accidents, eleven people lost their lives, and 183 were injured. The total direct financial losses amounted to 170 million €.

Fig. 1 Proportion of incidents in chemical plants with ageing as the cause (After [4]; MARS study)

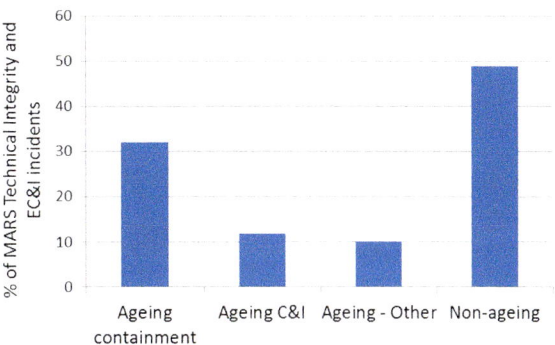

2.3 Nuclear Power Plants

More than in any other industrial sector ageing is a key issue in the nuclear industry. In the 1990s, the International Atomic Energy Agency (IAEA) started to develop a comprehensive set of publications on ageing management [5]. At that time, most existing nuclear power plants (NPPs) were 20–30 years old. These plants were then old enough to discover new ageing phenomena that slowly developed during the past decades. Moreover, new ageing phenomena were emerging as a result of more severe service conditions associated with increased plant performance, e.g. through implementation of the long-term operating experience obtained and/or the application of new technologies [5].

Proactive ageing management—Because of the serious consequences that may result from failure of components in a nuclear power plant, emphasis is, and should be, on *proactive* management of ageing. This in contrast with a 'run-to-failure' strategy, where components are replaced once they fail. The primary aim of proactive ageing- management programmes is to ensure the availability or required safety functions throughout the service life of the NPP. Moreover, effective management of ageing is also essential for achieving the desired plant performance and profitability of the plant.

The aim of physical or materials ageing management of structures, systems and components (SSC's), important for the safety of a plant, is to maintain their design safety margins above the SSC specific requirements, thus minimising the risks to which people and the environment are exposed. Based on experience, it is known that many SSC-failures are the result of ageing mechanisms, such as general and local corrosion, erosion-corrosion, radiation and thermal embrittlement, fatigue, creep, vibration and wear. To ensure a high level of plant safety, it is recommended to manage SSC-ageing effectively and proactively. This proactive ageing-management strategy includes both physical ageing and non-physical ageing that results from obsolescence, i.e. the plant's being out of date with respect to current safety regulation, standards, practices and technology.

Predictive models—The claim of being able to act proactively only holds if we know how long we avoid the moment that a material, structural component or system will fail. This presupposes that we have *predictive models* for the relevant ageing mechanisms and that we are able to monitor the progress of degradation with time reliably. According to the 2009 IAEA report Nr. 62 [5], the radiation embrittlement that leads to changes in bulk material properties has been successfully modelled. Also, the predictability of embrittlement of stainless steels and of irradiation and thermal degradation of polymers that are used in cable insulation and seals and that also produce changes in bulk-material properties, was considered adequate.

The IAEA report further states that the predictability of corrosion, wear and high-cycle fatigue, which produce changes at material surfaces and interfaces, is generally low. The resulting uncertainty has caused significant nuclear-power-plant unavailability and increased the costs for operation and maintenance. Important to mention is that operating experience in the nuclear industry has revealed degradation and failures of structures, systems and components caused by *previously unrecognized* ageing mechanisms. Given the potentially large consequences of NPP accidents, this is an important observation. The IAEA, therefore, states that in addition to improving the understanding and predictability of known ageing mechanisms, there is a need to provide for early detection of new ageing mechanisms. This requires sensitive and reliable monitoring and control devices and thorough understanding of the running processes in a plant.

In addition to what has been stated in the aforementioned IAEA report, the Fukushima disaster in 2011 has generated many new research questions, not only in Japan but also in Europe. The research focuses, among other things, on durable solutions for nuclear waste through immobilization and containment.

3 Ageing and Science

3.1 *Driving Forces Behind Ageing*

Ageing is often defined as a change of performance with elapsed time. This change of performance may refer to materials, components, structures or systems. How time per se can result in a change of performance is not easy to understand at first sight. How can a material or system 'at rest' change its performance? Maybe it is better to define time as *the domain* in which we describe observed changes in performance rather than *the cause* of these changes. But if we accept this, the question then still remains what the cause of changes in performance of a material, structure or system at rest could have been.

A closer look at any piece of matter 'at rest' tells us that the status of rest only applies to a certain length scale. Going down to the atomic scale, the world is in motion all the time. Fundamental entities, or *basic building blocks*, are continuously moving with a probability to leave one position for another one that fits them better. This phenomenon takes place in the time domain. It is an inherent feature of matter and lies at the basis of ageing processes.

On top of this inherent property of matter, we see, at different length scales, the existence of *gradients*. Gradients are driving forces, causing changes with elapsed time. At the boundary of any piece of material with its environment, gradients are present. These gradients concern, for example, temperature, humidity and radiation, and they may cause changes at the surface of the material. The chemical composition of the environment may cause changes at the surface with elapsed time, as well. Inside a material, clusters of atoms and molecules, e.g. crystals, are connected and build up a microstructure. At the contact points between crystals, however, atoms are liable to leave their position causing tiny changes in the performance of the material. In heterogeneous materials below a certain scale all materials are heterogeneous!—numerous interfaces offer sites for electrochemical activity, resulting in changes of the microstructure of a material with elapsed time and, hence, by ageing. Porous materials continuously communicate with their environment and never reach a condition of rest. This ongoing communication of porous materials, or better, porous systems, induces alternating stresses and strains in the system and gradually changes the micro- and nanostructure of the system and hence its performance.

The foregoing survey illustrates that a material 'at rest' is hardly conceivable. On smaller length scales, there is motion all the time, and a variety of driving forces, many of them in the form of gradients, promote the basic building blocks of a material to change their positions. This holds for all materials and systems. Basic building blocks search for a position where they feel energetically more comfortable. Their search will be more intensive the more they are forced to leave their positions for another one. This is the case particularly in made-made materials, where substantial *external* energy has been applied to let nature do what people want. But what we want is not always what nature wants. Nature is dictated by entropy laws—laws that *we* have to obey as well. By designing materials in a smart way, i.e. by minimizing internal gradients and concentrations of stresses and strains, there will be less reason for basic building blocks to leave their positions. Hence, the ageing process will slow down and the service life of materials, structures and systems will increase, at least from the perspective of ageing.

3.2 Change of Performance with Time

The early part of the lifetime of made-made materials, structures and systems is often characterized by a high probability of failure. It takes some time to overcome teething problems and to reach the required level of maturity and stability of the system. Once that point is reached, a stable period follows, until we arrive again at a period of increasing probability of failure. Exceeding a certain predefined probability of failure then marks the end of service life. The high probability of failure in

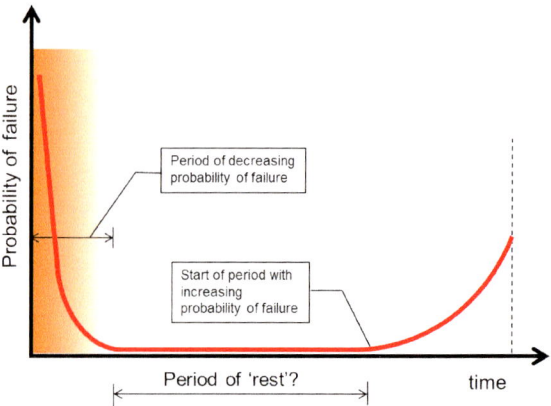

Fig. 2 Evolution of the probability of failure in a complex system, i.e. plant

the beginning, the subsequent period of 'rest' and the next period of increasing probability of failure is generally depicted with the bathtub curve (Fig. 2).

The length of the period with low probability of failure is of crucial importance for the economic performance of a system. The bathtub curve suggests that this is a period in which 'nothing happens'. It seems to be a period of rest, or dormant period. In the previous section, it has been indicated, however, that this 'macroscopic rest' does not mean that there is no activity, motion or change. Assuming that, during the period of low probability of failure, nothing happens is even misleading. This can be further illustrated by putting the bathtub curve of Fig. 2 upside down, as shown in Fig. 3. On the vertical axis, we now put 'Performance' instead of the 'Probability of Failure'. After a short period of teething problems, the system has reached the required level of performance. That is the level at which the system should demonstrate its capacity to meet safety and functional criteria, if possible, without intervention for maintenance or repair. It is the period of 'top-level sport' for all the atoms, molecules and interfaces of a material. When these basic building blocks give up and leave their positions, the period of decay starts. These first tiny

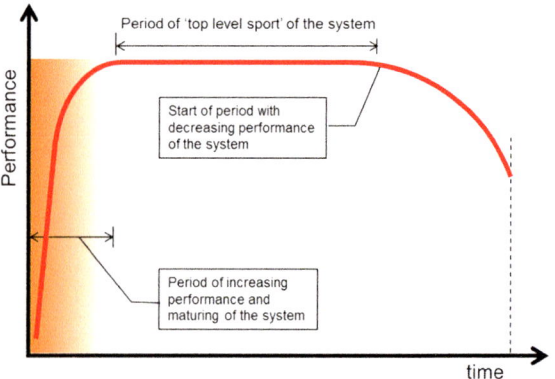

Fig. 3 Evolution of the performance of ageing materials, structures and systems

decay steps will most probably not be observed on the macroscale. The moment that the first basic building blocks give up doing their jobs can only be identified with extremely sensitive sensors in combination with comprehensive and appropriate material models. Here, chemistry, physics, electrochemistry, mechanics and mathematics *meet* each other and *need* each other for developing adequate tools for describing and predicting ageing processes at the fundamental level. In summary, and it is important to realise this!: ageing starts already in the stage *prior* to the moment that a system starts to change its performance!

4 Goals, Benefits and Investments

4.1 Mitigating Ecological Footprint

Research on ageing is supposed to result in tools for describing and predicting the (changes of) performance of a system with time. Moreover, by using these predictive tools, we should be able to design materials and structures such that their service life can be extended and savings can be harvested: savings not only in terms of money, but also in terms use of materials and energy and hence mitigation of the ecological footprint of production and construction processes. Whereas the *development* of man-made products generally results in an *increase* of the ecological footprint, *ageing studies* aim at *reducing* this footprint. Ageing studies are, therefore, all about *responsible stewardship* and must be considered as studies of the future!

4.2 Justification of Investment in Research on Ageing—the Global Picture

In Sect. 2.1, the value of the global infrastructure stock was estimated at US$53 trillion. Let us assume an average lifetime of these infrastructure assets of 50 years. Each year US$1.06 trillion has then to be spent on replacement of obsolete assets. Let us further assume that, through dedicated research, the average lifetime can be increased by 10%, i.e. from 50 to 55 years. The yearly replacement costs would then decrease from US$1.06 trillion to US$ 964 billion. This is a reduction in replacement costs of US$96 billion per year. Let us assume that, for saving these US$96 billion, we have to invest 20% of this amount in research, i.e. US$19 billion per year. Let us further assume that 50% of the required research money, i.e. US$9.5 billion, has to be spent on management-oriented research and the other 50% on science-oriented research. A part of this science-oriented research has to be spent on

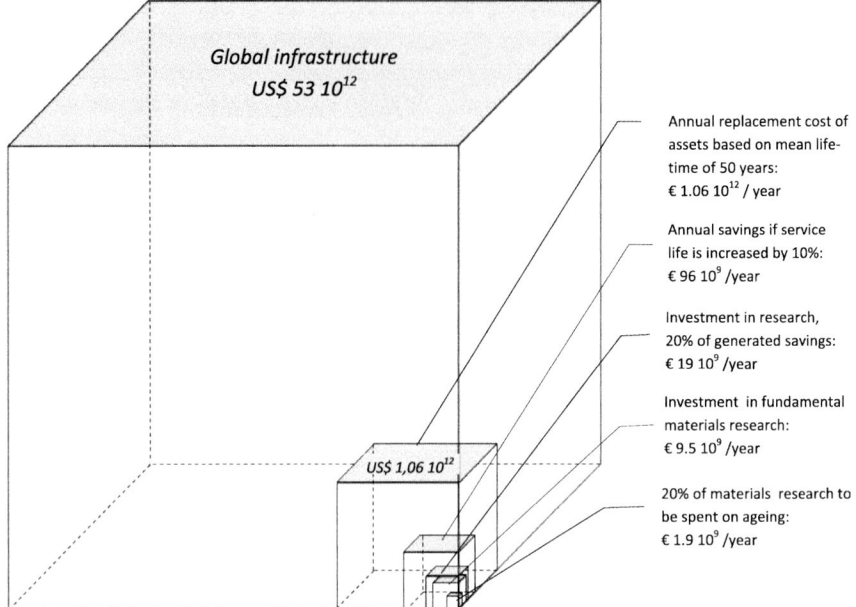

Fig. 4 Schematic presentation of required investment in ageing research for realising an extension of the mean service life of infrastructure of 10%. (interest/inflation not considered). Estimated average lifetime 50 years (after [6])

fundamental ageing research. A reasonable, though conservative, assumption is that 20% of science-oriented research, i.e. US$1.9 billion per year, should be spent on this fundamental ageing research. This amount of US$1.9 billion is 10% of the required research budget for realising the savings of yearly replacement costs (10% of US$19 billion) and only 2% of the targeted annual savings of US$96 billion. Schematically, this is shown in Fig. 4. By varying the assumptions in this exercise, other amounts for required investments are obtained, of course, but do not change the order of magnitude of these figures.

In this example, the percentages of 20% for investment in research needed to realise the targeted savings of 10% on the replacement costs and the percentage of 20% of the materials research for ageing studies were chosen more or less arbitrarily. Assuming that for realising savings an investment of only 20% of the targeted savings is needed is very optimistic. A percentage between 20 and 50% might be more realistic. Other percentages can be discussed as well. The overall concept, however, remains valid, as well as the conclusion that research finally pays off!

5 Concluding Comments

Ageing is a global issue: on a global scale, a trillion-dollar issue. Hence, solving ageing problems and mitigating the consequences of ageing is a global responsibility. Increasing awareness of the omnipresence of ageing and the recognition that ageing is a huge (financial) burden on the society and the environment forces people to look at their impressive scientific and technological achievements in a different way than usually done in the past.

Ageing is not an unwanted guest, but an inherent feature of nature that requires our full attention right from the beginning of any design activity. This is not easy, since ageing processes may start already in the stage that a structure is still performing according to its specifications! Defining ageing as a change of performance with time is, therefore, a bit misleading. Ageing starts already in the stage the structure is performing well. This makes ageing studies most challenging and requires ageing studies to be multi-scale and multi-disciplinary. An associated challenge is how to validate predictive ageing models when they apply to a period in which no macroscopic changes in performance are observed yet. Developing appropriate tests to accelerate fundamental ageing processes is badly needed.

Since ageing is a global issue, research projects on ageing should preferably be international. International platforms for in-depth discussions on ageing are considered essential for making progress in this challenging field. Those are the places where scientists, engineers, industry and governmental organisations should meet and define the research agenda on ageing. It's time to act. Think globally and, from this global perspective, act locally.

References

1. Long AE (2007) Sustainable bridges through innovative advances. Institution of civil engineers, presented at joint ICE and TRF fellows lecture p 23
2. Dobbs R et al (2013) Infrastructure productivity: how to save $ 1 trillion a year. McKinsey Global Institute, p 88
3. De Haan M et al (2009) The national capital of The Netherlands (in Dutch). In: De Nederlandse economie 2009, pp 129–140
4. Horrocks P et al (2010) Plant ageing study—phase 1. Health and Safety Executive, p 144
5. IAEA. (2009) Proactive management of ageing for nuclear power plants. Safety report series no. 62. International Atomic Energy Agency, p 83
6. van Breugel K (2015) Global dimension of ageing of infrastructure and our responsibility. 5th International Workshop on structural life management of power structures, Daejeon, Korea, Oct 2015, p 9

Part II
Fundamentals of Ageing

Investigation of Long-Term Behaviour of Elastomeric Seals for Transport and Storage Packages

A. Kömmling, D. Wolff and M. Jaunich

Abstract Elastomers are widely used as the main sealing materials in containers for low- and intermediate-level radioactive waste and as an additional component fo metal seals in spent-fuel and high-level waste containers. According to appropriate guidelines and regulations, safe enclosure of the radioactive container contents has to be guaranteed for lengthy storage periods of at least 40 years. Therefore, the understanding of seal ageing behaviour is of high importance and has to be considered with regard to possible dynamic events taking place during transport after storage. An accelerated ageing approach for compressed seals is presented, as well as some first results.

1 Introduction

Elastomers are widely used as main sealing materials in containers for low- and intermediate-level radioactive waste and as an additional component for metal seals in spent-fuel and high-level waste containers. This requires a service life in the range of several decades. According to appropriate guidelines and regulations, safe enclosure of the radioactive inventory has to be guaranteed for lengthy storage periods, as well as down to temperatures of −40 °C for transportation. Therefore, the understanding of seal behaviour in general is of high importance, and especially ageing of elastomeric seals has to be considered. Possible dynamic loads may occur during the whole interim storage period (so far approved in Germany for up to 40 years) and during transportation after storage. To fulfil their purpose, the seals have to remain in good shape with adequate resilience to ensure a certain contact pressure and without, e.g. cracks crossing the sealing surface that would act as leakage paths.

A. Kömmling · D. Wolff · M. Jaunich (✉)
Division 3.4 Safety of Storage Containers, BAM Federal Institute
for Materials Research and Testing, Berlin, Germany
e-mail: matthias.jaunich@bam.de

© Springer International Publishing AG 2018
K. van Breugel et al. (eds.), *The Ageing of Materials and Structures*,
https://doi.org/10.1007/978-3-319-70194-3_2

Ageing of materials is an undesirable, but unavoidable, process that can lead to (non-reversible) changes in, e.g. mechanical properties, thermal properties, colour and chemical composition [1]. The origin of these property changes can be innate to the material or caused by influences from the environment. Typical innate effects are, e.g. trapped stresses or remaining crosslinking agents. Effects caused by the environment are, e.g. oxidation processes, influences from light and other radiation sources and heat.

The changes of the material properties can be caused by purely physical or chemical processes. Physical ageing can take place as molecular rearrangement, demixing or crystal growth. Usually it is driven by thermodynamic non-equilibrium conditions, which are often innate effects. Chemical ageing results in changes of the chemical composition of the material by chemical reactions. Ageing of polymeric materials is generally considered to consist of two pathways. One is the formation of additional crosslinks, and the other is chain scission [2]. Generally, these two processes take place in parallel while one process is dominant. This means that some materials primarily tend to crosslinking and others to chain scission. The predominant mechanism can change over the whole ageing period. Additionally, temperature has an important influence on the ageing process. It is possible that a material forms a high amount of crosslinks at one temperature and at, e.g. that, at a higher temperature, chain scission becomes dominant.

To describe the effect of ageing on the performance of a material, it is important to test the relevant properties. For long-term applications, an accelerated test is required to ensure long-term safety. This approach is described by standards, e.g. ISO 188 [3]. Often these standards address only the general approach for accelerated ageing and assume an Arrhenius-like behaviour of the ageing process, whereas the necessary prerequisites and the data analysis are often not described in detail. Moreover, often these standards describe ageing of standard samples and not of components. The samples are often unstressed and therefore the effect of a continuous deformation of, e.g. an elastomeric seal cannot be evaluated even if such permanent stress might lead to different ageing effects as, e.g. predominantly chain scission of stressed chain molecules.

In the literature, there are several examples that show non-Arrhenius ageing effects and, therefore, the need for a more detailed analysis of the test results and a careful selection of test parameters [4, 5]. Especially the interactions in complex technical elastomer compounds consisting of a multitude of different components are hard to predict.

As some material properties are more sensitive to changes during ageing than others, it is important to measure a range of various properties, e.g. chemical, mechanical and thermal properties. However, some measured properties, such as tear strength, have typically very little or a highly complex correlation with sealing performance [6]. Other standard methods such as compression set or stress relaxation measurement are more related to seal applications, providing an indirect value for seal performance [5]. An overall goal of ageing investigation of seal materials is to determine which properties are sensitive to occurring changes and how they are correlated with the safety-relevant performance of the seal.

Our previous work was concerned mainly with the low-temperature behaviour of elastomeric seals and modified compression set tests [6–9].

In this paper, we focus on our approach to investigate the behaviour of elastomeric seals over extended periods of time at elevated temperatures. We describe our approach to investigate the occurring changes of material properties of elastomeric seals by ageing performed at various temperatures and under assembly conditions. First results are presented as well. One goal is to determine the influence of compression during the ageing period, and therefore compressed and uncompressed samples are aged and compared. The temperature influence must also be analysed to ensure an appropriate acceleration of the ageing processes (e.g. exclusion of additional sample degradation as a result of high ageing temperatures).

2   Material Selection and Ageing Conditions

As mentioned, the use of elastomeric seals in containers for radioactive waste is our motivation to start the described research, and, hence, the investigation parameters and the material selection are related to this application.

The material selection focuses on material classes that are internationally used for this application. Therefore, a fluorocarbon rubber (FKM) and an ethylene-propylene-diene rubber (EPDM) were selected. Additionally hydrated nitrile butadiene rubber (HNBR) was tested for comparison.

The seal dimensions were chosen with a rather high torus diameter of 10 mm and an inner diameter of 190 mm to have a sufficient amount of sample material for the various analytical methods to be performed. Mechanical ageing conditions should also be varied to investigate whether a compressed seal shows a different ageing behaviour in comparison to an uncompressed sample.

Ageing temperatures are selected in the range from 75 °C up to 150 °C and are therefore spread broad enough to have a significant acceleration. With a total of four test temperatures, a sufficient amount of data should be available to judge the applicability of typical time-temperature-superposition approaches as, e.g. the Williams-Landel-Ferry approach combined with the Arrhenius approach.

During ageing, samples are stored in unstressed condition as complete O-rings and in compressed condition with a degree of compression of 25%. To compress the seals, they are positioned between two metal plates that are screwed together until both plates are pressed against distant pieces. The devices used are schematically shown in Figs. 1 and 2.

Besides these two ageing conditions, it is important to allow a correlation of measured changes in material properties with seal function, which is possible with seals mounted in flanges. These flanges are pictured in Fig. 3. This setup enables leakage-rate measurements of mounted seals by a pressure-rise measurement after

Fig. 1 Uncompressed
O-rings ageing on racks of
punched sheets

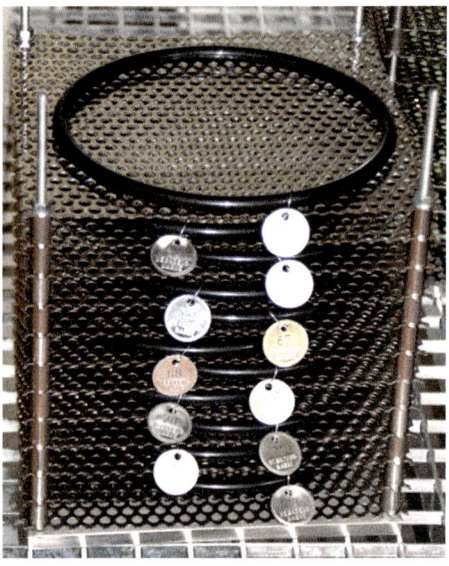

Fig. 2 O-rings ageing in
compression between plates

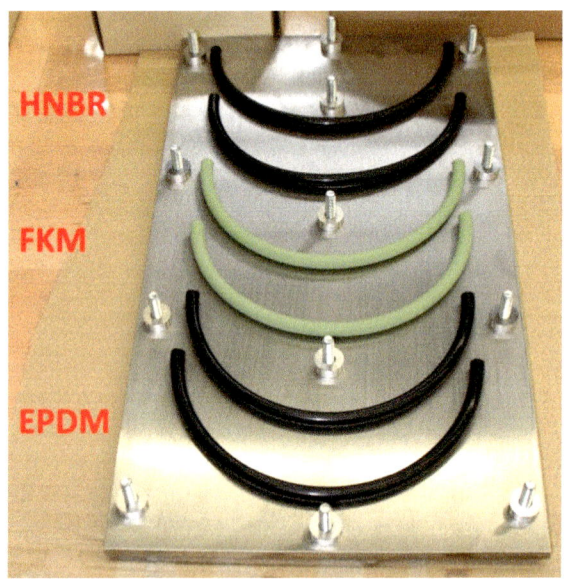

predefined ageing times. The principle setup for such a measurement is described in
[6, 10].

The samples are removed from the high-temperature chamber at predefined
times to be investigated by several methods. The ageing times are given in Table 1.

Fig. 3 Test flanges for leakage/rate measurements

Table 1 Ageing times and number of O-rings for the various ageing setups

Storage time	0 h	3 h	10 h	1 d	3 d	10 d	30 d	100 d	0.5a	1a	2a
Uncompressed samples	3	1	1	1	1	1	1	1	1	1	1
Compressed samples		1	1	1	1	1	1	1	1	1	1
Samples in flange	3										

3 Analytical Methods

For the investigation of ageing effects, several methods are helpful to describe the occurring changes and to quantify the influence on the seal performance. For the investigation of structural/morphological changes, suitable methods are, e.g.

- FT-IR infrared spectroscopy for detecting changes in the chemical structure, e.g. crosslinking or insertion of oxygen (oxidation)
- Differential Scanning Calorimetry (DSC) for detecting phase transitions over temperature
- Thermal Gravimetric Analysis (TGA) for measuring the mass loss of a sample during a heating programme as a result of decomposition processes.

For investigation of mechanical properties, several methods can be applied, such as

- Dynamic Mechanical Analysis (DMA) for measuring the modulus in dependence of temperature to characterise the viscoelastic properties of the material and occurring transitions
- Compression Set (CS) as a seal-related property that measures the recovery behaviour after compression at various temperatures

- Compression stress relaxation as a measure of sealing-force decay
- Hardness
- Tensile testing.

The observed changes shall be correlated with the leakage rate of the aged seals.

4   First Results and Discussion

Results of compression stress relaxation experiments of the elastomers are shown in Fig. 4 where the decrease of sealing force F is normalized to the initial force F_0. Three samples of each material were tested at 150 °C [11,12].

After the initial fast drop within the 1 h of the experiment, the materials show a distinct behaviour. FKM shows the slowest decrease with time. HNBR decreases fast within the first days, but the decrease rate becomes smaller and the compression force of EPDM continues to decrease rapidly. This is probably due to diffusion-limited oxidation (DLO) effects [13, 11] which can appear in thick samples at high ageing temperatures. Then, the ageing reactions are so that fast on the outer sample side, which is exposed to air containing oxygen, that it consumes most of the oxygen or forms an oxygen-diffusion barrier through crosslinking reactions (indicated by an increase in hardness). Thus, the inner part of the sample is protected from oxidation and can retain its elastic properties. As the micro hardness measured across the O-ring cross-section (shown in Figs. 5 and 6) indicates, these heterogeneous ageing effects occur in HNBR after 30 d at 150 °C, but not in EPDM. Thus, EPDM degrades across the whole sample, while HNBR can retain an elastic core.

Results of hardness measurements on the sealing area of HNBR seals are shown in Figs. 7 (uncompressed) and 8 (compressed). The uncompressed samples show a

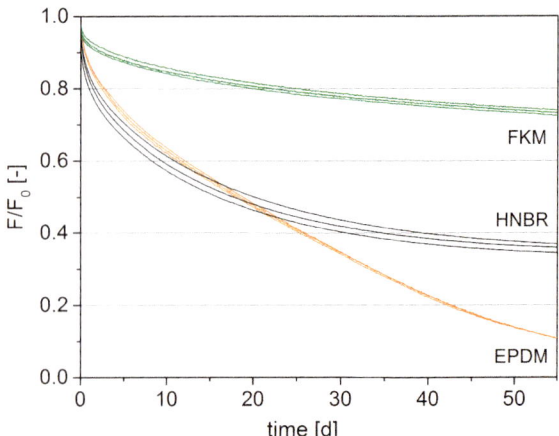

Fig. 4 Decrease of normalized sealing force vs. time at 150 °C

Fig. 5 Micro hardness profiles across HNBR seals (Ø 10 mm) aged for 30 d

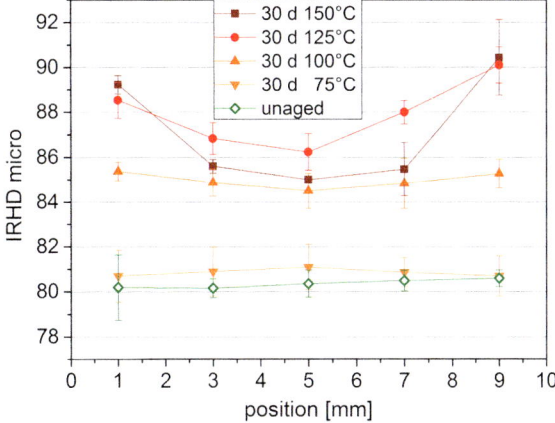

Fig. 6 Micro hardness profiles across EPDM seals (Ø 10 mm) aged for 30 d

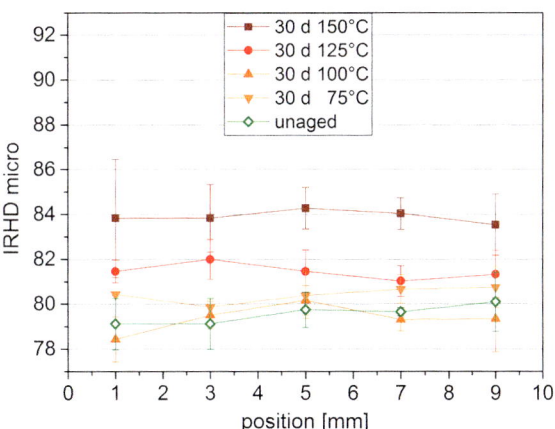

Fig. 7 Hardness measured on the sealing area of uncompressed HNBR seals

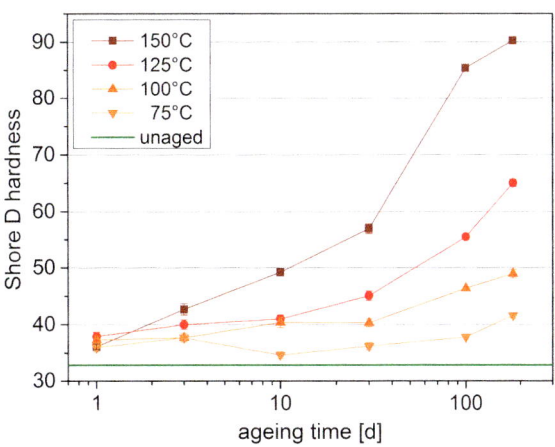

Fig. 8 Hardness measured on the sealing area of compressed HNBR gaskets

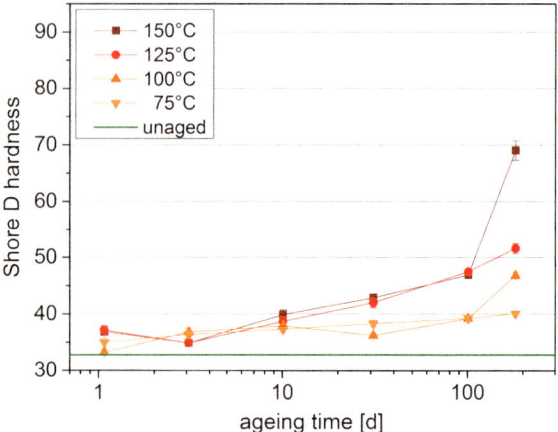

higher hardness increase compared to the compressed samples. This is because the sealing area of the compressed seal is covered by flanges. Thus, oxygen access is hindered for this sample arrangement and ageing is slowed. This illustrates that the reduced area for oxygen access has to be taken into account for seals aged in compression.

5 Conclusion

The investigation of the ageing behaviour of elastomeric seals is an important factor for the assessment of the long-term performance of sealed containers.

Ageing is a complex process that involves multiple reactions. To describe the material behaviour in the necessary extent, an in-depth understanding of these reactions and their influence on the sealing function is necessary. As standard procedures of polymer ageing are typically designed for the lifetime estimation for goods having much shorter service lives, those standard regulations seem not well-fitted for the lifetime estimation over several decades. Therefore, we propose and perform ageing experiments tailored to the particular needs for elastomeric-seal evaluation with the focus on application for a lifetime of several decades. First results have shown that the access of oxygen plays an important role in the degradation, which is influenced both by the dimensions of the sample (thickness) and by the compression assembly (covering by flanges).

Acknowledgements The authors are grateful for the support by our colleagues M. Goral, S. Schulz and L. Qiao.

References

1. DIN 50035:2012–09, Begriffe auf dem Gebiet der Alterung von Materialien—Polymere Werkstoffe
2. Ehrenstein GW, Pongratz S (2007) Beständigkeit von Kunststoffen, vol 1. Carl Hanser Verlag, München
3. ISO 188:2011, Rubber, vulcanized or thermoplastic- Accelerated ageing and heat resistence tests
4. Celina M, Gillen KT, Assink RA (2005) Accelerated aging and lifetime prediction: review of non-Arrhenius behaviour due to two competing processes. Polym Degrad Stab 90:395–404
5. Marlier R, Andre R, Malesys P, Issard H (2006) Seal life of EPDM O-rings at high temperature determined by unique method. Packag Trans Storage Secur Radioact Mater 17:57–62
6. Jaunich M (2012) Tieftemperaturverhalten von Elastomeren im Dichtungseinsatz. Doktorarbeit an der Bundesanstalt für Materialforschung und Prüfung, Berlin
7. Jaunich M, Stark W, Wolff D (2010) A new method to evaluate the low temperature function of rubber sealing materials. Polym Test 29:815–823
8. Jaunich M, Stark W, Wolff D (2012) Comparison of low temperature properties of different elastomer materials investigated by a new method for compression set measurement. Polym Test 31:987–992
9. Jaunich M, Wolff D, Stark W (2013) Low temperature properties of rubber seals—results of component tests. KGK-Kautsch. Gummi Kunstst 66:26–30
10. Jaunich M, von der Ehe K, Wolff D, Völzke H, Stark W (2011) Understanding low temperature properties of elastomer seals. Packag Trans Storage Secur Radioact Mater (RAMTRANS) 22:83–88 (2011)
11. Kömmling A, Jaunich M, Wolff D (2016) Effects of heterogeneous aging in compressed HNBR and EPDM O-ring seals. Polym Degrad Stab 126:39–46
12. Kömmling A, Jaunich M, Wolff D (2016) Ageing of HNBR, EPDM and FKM O-rings, KGK 69(4):36
13. Wise J, Gillen KT, Clough RL (1997) Quantitative model for the time development of diffusion-limited oxidation profiles. Polymer 38:1929–1944

What Can Be Learnt from Ageing in Biology and Damage-Tolerant Biological Structures for Long-Lasting Biomimetic Materials?

Thomas Speck, Marc Thielen and Olga Speck

Abstract Ageing in biology and the principles of how living beings deal with shortcomings occurring during ageing on the structural and functional level are exemplified. The presented examples mainly focus on damage repair and damage tolerance in biological materials and structures, and on what can be learnt for ageing man-made materials and structures by using bio-inspired approaches. The potential of such an approach is specified by three examples comprising biomimetic self-repairing foam coatings for pneumatic structures, bio-inspired self-healing elastomers and biomimetic damage-tolerant fibre-reinforced gradient foams with high-energy dissipation.

Keywords Ageing in biology · Biomimetic materials · Self-healing
Long-lasting · Energy-dissipation · Damage tolerance

1 Introduction

Ageing and eventually death are central paradigms of life. They are part of everybody's life, which makes us genuinely interested in these processes. Only unicellular organisms, as, for example, bacteria and some algae, can be considered as potentially immortal as they multiply via cell division. This asexual way of

T. Speck (✉) · M. Thielen · O. Speck
Botany: Functional Morphology & Biomimetics, Faculty of Biology,
Botanic Garden of the University of Freiburg, Freiburg im Breisgau, Germany
e-mail: thomas.speck@biologie.uni-freiburg.de

T. Speck · M. Thielen
Freiburg Materials Research Center (FMF), Freiburg im Breisgau, Germany

T. Speck · O. Speck
Freiburg Center for Interactive Materials and Bioinspired Technologies (FIT),
Freiburg im Breisgau, Germany

T. Speck · O. Speck
Networks of Competence Biomimetics, Baden-Württemberg, Germany

© Springer International Publishing AG 2018
K. van Breugel et al. (eds.), *The Ageing of Materials and Structures*,
https://doi.org/10.1007/978-3-319-70194-3_3

27

reproduction has the effect that after each division life starts newly for both daughter cells, which are identical to the mother cell. As soon as non-unicellular organisms with specialized cells began to evolved, they became mortal. *Volvox* is a spherical colonial green alga, up to one millimetre in diameter, having specialised cells for motion and photosynthesis (vegetative cells), as well as for propagation (generative cells). This green alga is often cited as being an example of how organisms may have looked which mandatorily die (after sexual propagation) and which are considered to have evolved the first carcass in life history. Having this in mind, one could argue that specialisation and thus more complex body plans, inevitably entail ageing and eventually death [1, 2].

2 Ageing and Damage "Repair" in Biology

Ageing in living beings takes place on various hierarchical levels ranging from the molecular (e.g. ageing of the DNA) and the ultrastructural level, the cellular and tissue level to the level of organs and finally the entire organism. Ageing on the tissue and organ level is often linked to wear, damage, injury, failure and, concomitantly, a reduction of functionality and finally entire loss of function (Fig. 1). As some plants and animals can become comparatively old, these effects lead to increasing problems for these living beings with advancing age. During evolution, various principles have evolved to deal with these "unwelcome effects" of ageing. As basic "method" to cope with change—either caused by ageing processes and/or by other internal (growth stresses, damage due to overloading of tissues and organs etc.) or external factors (predators, environmental impacts as e.g. wind loads or rock falls etc.)—various, often highly efficient, self-repair mechanisms have evolved. These mechanisms act on all hierarchical levels of living beings and include various mechanisms, such as DNA-self-repair (e.g. after UV-induced damages) or the

Fig. 1 a Breakage of a large branch in the lowland tropical rain forest of French Guyana. **b** Stem failure in a pine tree as a consequence of overcritical wind loads during a storm, south of France. **c**, Longitudinal split in a large stem of an old oriental beech (*Fagus orientalis*) caused by overcritical wind loads, Botanic Garden Freiburg © Plant Biomechanics Group Freiburg

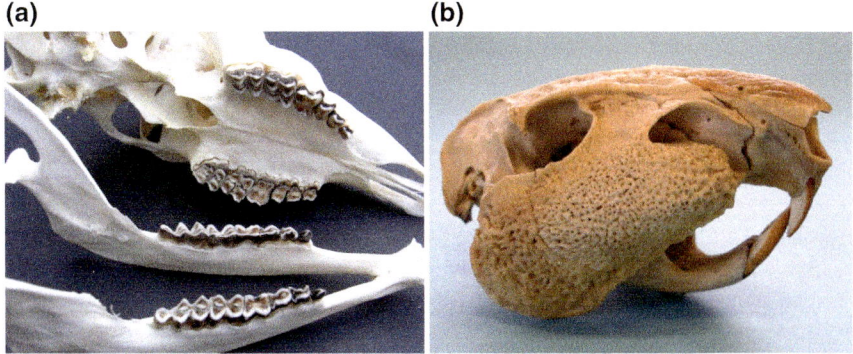

Fig. 2 a Skull of a 13-year-old chamois (*Rupicapra rupicara*) with markedly worn, barely functional teeth. **b** Skull of a beaver (*Castor fiber*) showing the self-sharpening teeth found in many rodents © Plant Biomechanics Group Freiburg

partial, or even entire, regeneration of limbs as also observed in some vertebrates, e.g. various amphibians [1, 2]. Another self-repair mechanism to ensure survival in all "higher" animals is blood coagulation. Without the latter mechanism, even small internal or external wounds would be lethal to the organism. The effect of non- or malfunctioning blood coagulation is well know from haemophiliac persons who can only survive due to medications substituting missing coagulation factors.

Another mechanism found in animals to deal with wear and ageing is the replacement or restoration of lost or worn tissues or organs. Some of the most well-known examples are linked to teeth that are exposed to high mechanical loads during prey capturing, as well as abrasion associated with masticatory processing, and thus are essential for survival. Examples include the continuous tooth-replacement taking place in sharks (some shark species shed and replace more than 30,000 teeth in a lifetime), the horizontal tooth replacement found in elephants, or the continuously growing and self-sharpening teeth of rodents (Fig. 2) [1, 2]. Also, plants exhibit a great variety of self-repair mechanisms for both external and internal fissures and wounds. Examples are wound-callus formation (seen, e.g. in woody plants if a branch is broken or cut off and the wound is closed), wound sealing due to latex coagulation, closure of fissures by deformation of the organ or swelling of pressurized cells into fissures that were caused by internal growth processes [3, 4]. Similarly as in animals, replacement of entire organs and tissues also evolved in plants as a common mechanism to deal with wear or damage. This is facilitated in plants due to the fact that plants typically have a modular body plan with many reiteratively arranged substructures (leaves, branches, twigs, roots) which can easily be replaced and re-grown [2].

In spite of all these more or less sophisticated mechanisms that have evolved over a time period of 3.8 billion years, all "higher" organisms have a finite lifespan and ultimately die. As known from demography, the absolute maximum age characteristically outdates the "typically" mean maximum age by far. The record

holder in animals is the black clam (*Arctica islandica*) with over 500 years, followed
by the greenland shark (*Somniosus microcephalus*) with a maximum age between 300
and 400 years and the Aldabra giant tortoise (*Aldabrachelys gigantea*) with more than
250 (perhaps 300) years. Humans can become over 120 years old with an actual
proven "record" of 122 years. Plants often become much older. Record holders in the
plant kingdom are bristlecone pines (*Pinus longaeva*) with an absolute maximum age
of over 5,000 years for individual trees, and giant redwoods (*Sequoiadendron
giganteum*) with an absolute maximum age of up to 4,000 years (Fig. 3). The latter
species, attaining a height of over 120 m and a stem diameter of more than 11 m, is
also one of the largest trees and even (individual) organisms that have ever existed. It is
worth mentioning that these maximum ages relate to "individualized" organisms. In
the last decade, it was found that clonal plants and animals may become significantly
older, if the age of the clone is considered as the "lifetime" of a genetic organism (some
of these clonal organisms also belong to the largest living beings known). An age of
about 80,000 years has been attributed to a clonal quaking aspen colony (*Populus
tremuloides*) in Utah. For some reef-forming deep-sea corals, a lifetime of several
thousand years is extrapolated based on recent growth rates, and for the arctic giant
sponge (*Anoxycalyx joubini*) and arctic glass sponges an age of 10,000 years was
assumed, an estimation which, however, is the subject of an ongoing debate. [5, 6 and
www.wikipedia.org, approached on 2017-10-12].

(a)   **(b)**

Fig. 3 a Giant redwood (*Sequiadendron giganteum*), the picture shows the "General Sherman"
tree, the biggest tree as to volume (1,487 m^3) and mass (1256 t), which has an age of ca.
2,200 years, a height of about 84 m and a basal-stem circumference of more than 31 m.
b Bristlecone pine (*Pinus longaeva*) in the Botanic Garden Freiburg © Plant Biomechanics Group
Freiburg

3 Bio-Inspired Damage "Repair" and Damage Tolerance in Ageing Technical Materials and Structures

The following examples are presented to demonstrate what can be learnt from biology to overcome problems caused by ageing in technical materials and structures, and how damage tolerance and damage repair in man-made structures can be inspired by biology.

4 Biomimetic Materials and Structures

Biological "constructions" often possess outstanding mechanical properties that are mainly based on a complex hierarchical structuring, including a multitude of interfaces on various structural levels. They are not built of a vast variety of constitutive materials, as opposed to many traditionally engineered structures, but are characterized by a limited number of basic chemical components and a wide diversity of micro- and nanostructures [7, 8]. The extremely efficient biological "materials design" is brought about by the evolution of hierarchical structures covering more than ten orders of magnitude and being well adapted to the requirements at each level of hierarchy. Many biological materials possess, in addition to their fascinating mechanical functions, "self-x-properties", such as self-organization, self-cleaning and/or self-healing. The combination of the respective properties enable them to interact very efficiently with their respective environment and to delay "symptoms" of ageing. These biological solutions are cost- and energy-efficient, multi-functional, long-lasting and environmentally friendly.

During the last decade, the development of novel sophisticated methods made possible quantitative analyses and simulations of the form-structure-functions-relationship on various hierarchical levels and thus revealed new fascinating insights into multi-scale mechanics and other functions of biological materials. This deepened the understanding of how biological materials cope with ageing problems and/or damage caused by internal or external factors. Furthermore, new production methods enable, for the first time, the transfer of these outstanding biological properties and mechanical principles to innovative biomimetic products [3, 4, 7].

Based on two current R&D projects of the Plant Biomechanics Group Freiburg, serving as examples for a successful transfer of ideas from biological concept generators, the development of hierarchically organized biomimetic self-repairing and highly damping damage-tolerant materials and structures are presented. These examples also demonstrate how biomimetics may help to deal with problems occurring and accumulating during the ageing of technical materials and structures.

5 Biomimetic Self-Repairing Materials and Structures

In biological self-repair, typically two phases can be discerned: a fast, self-sealing process restoring functionality and a slower, self-healing process reconstituting structural integrity [3, 4, 9]. Role models for the development of a self-repairing foam coating for pneumatic structures were pipevines of the genus *Aristolochia*. In these plants, internal fissures occur in the peripheral ring of lignified stabilizing tissue caused by radial and tangential stresses. These stresses are evoked by high strains resulting from secondary growth processes of the internally located wood and phloem tissues. As soon as such a fissure occur, cells of the outer cortex, which are under internal pressure (turgor), swell into these fissures and seal them very efficiently [3, 4, 10]. Inspired by the self-sealing processes analysed in the stems of the pipevines, a self-repairing foam-coating for pneumatic structures was developed. This biomimetic foam coating is based on closed-celled polyurethane foams and allows for sealing holes of up to 5-mm diameter in the membranes of pneumatic structures. Best results are obtained when curing of the foam under an overpressure of 1–2 bar (Fig. 4) is performed. This procedure leads to a thin, pre-strained, self-repairing foam coating (thickness of a few millimetres), which is located at the inner surface of the membrane. As a consequence of the pre-straining and of the compression strains occurring due to the arched shape of pneumatic structures, the inlying foam layer relaxes within the hole and seals it. By this process, holes that are either caused by damage or by ageing processes of the membrane can be sealed entirely, or at least the air leakage in pneumatic structures, for example, rubber boats or tensairity® buildings, can be reduced by a factor of up to 1,000, [3, 10, 11].

A second example are bio-inspired self-repairing elastomers for technical seals and dampers. These elastomeric structures are typically exposed to high cyclic loading and often fail suddenly due to (uncontrollable) growth of microcracks, long before the actual loading limit of the material is reached. This type of failure can be considered as a consequence of an age-induced accumulation and/or increase of microcracks. The basic lesson from biology in this project was to accept microcracks as unavoidable material imperfections and attempting to stop the growth of the ubiquitous microcracks before they reached a critical length. The biomimetic solution in this case was inspired by self-sealing and self-healing processes found in latex bearing plants, e.g. the weeping fig (*Ficus benjamina*) and spurges (*Euphorbia* spp.), which very efficiently seal micro- and macrofissures by latex coagulation [4, 12–14]. Inspired by the results of analyses of self-sealing processes in the biological role models, a self-healing elastomer was developed based on microphase-separated polybutadiene rubber (PBR)/hyperbranched polyethyleneimine (PEI) blends. The self-healing efficiency was measured after cutting a PBR/PEI sample in half, re-joining it under compression and annealing it for 12 h at 100 °C, and then storing it for another 12 h at room temperature. A self-healing efficiency of up to 44% was found for tensile strength (strain rate: 50 mm min^{-1}), which is very promising for further research and improvement [3, 4, 15].

Fig. 4 **a** Habitus and stem cross-section with repaired fissures in the outer strengthening tissue ring of the twining pipevine *Aristolochia marcophylla* which served as a role model for self-repairing foam coatings for pneumatic systems. **b** Custom-made set-up for testing and comparing the repair efficacy of coated and uncoated membranes © Plant Biomechanics Group Freiburg

6   Biomimetic Damage-Tolerant and Energy-Dissipating Materials and Structures

A high damage tolerance and energy absorption can be found in some fruit peels and in the bark of some trees and lianas. In the case of fruits, the external structure (peel) protects the inner parts (seeds and fruit pulp) that are vital for germination, reproduction and propagation against impact and splitting open when hitting the ground after being shed. In the case of tree trunks and liana stems, the inlaying living tissues that are essential for vital physiological processes, as e.g. secondary growth, are protected by a thick, sometimes spongy, bark against rock falls (some alpine trees) or against impingement upon the host trees (some lianas). These energy-dissipating envelopes make these plant organs damage-tolerant during ripening and growth (i.e. during ageing), and protect them during the lifespan of the respective organ. Especially in trees and lianas, the prevention from damage by these external protection layers are—together with effective, self-repair processes in case damage could not be avoided—key factors for longevity and preservation of functionality for up to several thousand years.

The pomelo (*Citrus maxima*) is the largest and heaviest fruit of the genus *Citrus* (mass typically 1–2 kg, however sometimes up to 6 kg) and grows on trees up to 15-m tall. By studying the fruit peel and the damping properties of pomelo fruits, the hierarchical structuring of the fruit peel could be identified as the key element for the protective and shock-absorbing function. Due to this hierarchical structuring, the fruits can withstand drops from up 10 m onto a concrete floor without visible damage, and an impact energy of more than 80 J can be dissipated [16, 17]. In pomelo fruits, up to seven different hierarchical levels of fruit and fruit peel can be discerned (Fig. 5). The fruit peel mainly consists of a lightweight open-pored foamy tissue whose specific density gradually changes from the epidermis (outside) towards the pulp (fruit flesh), and which is penetrated by a three-dimensional network of vascular bundles. The struts of the foam comprise living parenchyma cells which are approximately cylindrical in shape. Our results show that this gradient foam structure can very effectively damp mechanical impacts. This is further improved by the three-dimensional network of bundles which helps to distribute local impact forces occurring at the contact region with the soil to a global straining involving a huge amount of the peel volume. By this structure, over 90% of the kinetic energy are dissipated in free-fall tests with pomelo fruits [16, 18]. This excellent damping behaviour could be confirmed by drop-weight tests with peel samples in which also the influence of the hydration state of the peel on the energy dissipation was analysed [18, 19].

Based on an abstraction of the results for the pomelo peel, biomimetic, shock-absorbing, fibre-reinforced gradient foams were developed [17, 20]. Tests with these light-weight biomimetic foams show that they also have an excellent damping and energy-dissipation behaviour. As the abstraction proved that the capability for energy dissipation is (mainly) caused by the hierarchical structuring of the gradient fibre-reinforced biological foam [16, 18], the technical foam

Fig. 5 Various hierarchical
levels of the pomelo fruit and
fruit peel (*Citrus maxima*) ©
Plant Biomechanics Group
Freiburg

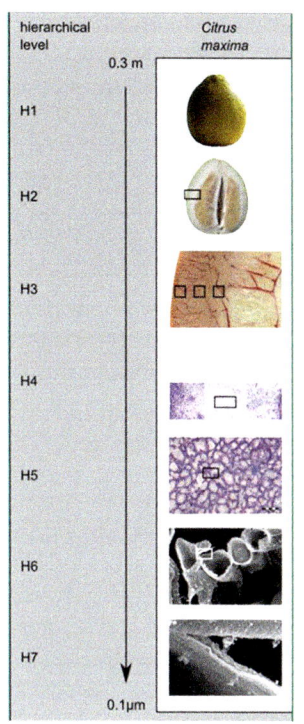

structures can be produced by using various material groups, e.g. metals or polymers, and can be used for various types of technical protection structures, including crash boxes for cars, transport vessels for hazardous goods or protective wear and helmets for cycling or sports [7, 17].

Bio-inspired energy dissipating "envelopes" may help to protect man-made structures from external damage that accumulates during a lifetime, and therefore may protect essential central functions in ageing materials and structures against loss of function and breakdown.

7 Discussion

The short review just presented, shows how some selected living beings cope with damage and wear problems caused by ageing or other internal and external influences. It offers at least four lessons that can be learnt and can be used for the development of bio-inspired materials and structures that may help with (mechanical) problems occurring during ageing in technical materials:

(1) The ability to self-repair is a central property of living beings as it helps to preserve functionality of vital functions during ageing and/or after damage.

Biological self-repair occurs on all hierarchical levels and takes advantage of the interfaces for sealing or keeping microfissures small and below critical levels. Bio-inspired self-repair therefore may help to increase service life in ageing technical materials and structures [3, 4].

(2) Due to their internal (hierarchical) composition, biological materials and structures are often highly damage tolerant and additionally show a "fail-safe" mechanism. The latter means that these materials and structures have a distinctly benign fracture behaviour, and that final break down is "announced" and often delayed by so-called pre-failure events [7, 21, 22].

(3) A third lesson that can be learnt for bio-inspired technical applications can be named "replacement in action": worn or otherwise damaged structures are replaced continuously during full functionality of the system acting at a higher level, i.e. without stopping the system for the replacement process. This is, for example, the case in tooth replacement in elephants and sharks.

(4) A last lesson to be learnt is the fact that life span in biology is limited and that with ageing the performance decreases in many aspects. Therefore, life and ageing in biology can, simplistically, be seen as a permanent fight against the accumulation of local failures and repair of damage in order to keep the functionality on a sufficient level to survive. In engineering and especially in architecture, humans still try to construct for "eternal use and functioning". Biology teaches us that this attitude has to be reconsidered. A limited life span of technical products and buildings, adapted to the respective application, scheduled during construction, communicated to the customer, and ideally coupled with recyclability of all material components, may often be the superior solution. Such a limited lifetime could help to significantly decrease problems with ageing of technical materials [7].

Acknowledgements We thank our colleagues and their research groups for excellent cooperation during the projects presented above, especially Andreas Bührig-Polaczek (RWTH Aachen), Claudia Fleck (TU Berlin), Rolf Luchsinger (EMPA Dübendorf), Rolf Mülhaupt (University of Freiburg and FMF) and Anke Nellessen (Fraunhofer Institute UMSICHT Oberhausen). We also thank Martin Köhn (University of Freiburg) for the permission to use his skull collection for making pictures. For financial support, we are grateful to (1) the German Research Foundation (DFG) for funding within the priority program 1420 "Biomimetic Materials Research: Functionality by Hierarchical Structuring of Materials" and within the Transregional Collaborative Research Centre 141 "Biological Design and Integrative Structures", (2) to the German Federal Ministry of Education and Research for funding within the scope of the programme BIONA and the within the scope of the funding programme "Ideenwettbewerb: Bionik—Innovationen aus der Natur", and (3) to the Ministry of Science, Research and the Arts of Baden-Württemberg in the framework of the "Sustainability Center Freiburg".

References

1. Sadava D, Orians G, Heller HC, Hillis D, Berenbaum MR, (Markel J ed.) (2012) Purves Biologie, 9th ed. Springer Spektrum, Heidelberg
2. Lexikon der Biologie (1999–2004) Elsevier/ Spektrum Akademischer Verlag, München

3. Speck T, Bauer G, Flues F, Oelker K, Rampf M, Schüssele AC, Tapavicza v. M, Bertling J, Luchsinger R, Nellesen A, Schmidt AM, Mülhaupt R, Speck O (2013) Bio-inspired self-healing materials. In: Fratzl P, Dunlop JWC, Weinkamer R (eds) Materials design inspired by nature: function through inner architecture, RSC smart materials No. 4, The Royal Chemical Society, London, pp 359–389

4. Speck T, Mülhaupt R, Speck O (2013) Self-healing in plants as bio-inspiration for self-repairing polymers. In: Binder W (ed) Self-healing materials. Wiley-VCH, Weinheim, pp 69–97

5. Flindt R (2003) Biologie in Zahlen, 6th edn. Spektrum Akademischer Verlag/Gustav Fischer Verlag, Heidelberg

6. Speck T, Schmitt M (1992) Tabellen. In: Schmitt M (ed) Biologie im Überblick, Lexikon der Biologie, vol 10. Herder Verlag, Freiburg, pp 187–328

7. Masselter T, Barthlott W, Bauer G, Bertling J, Cichy F, Ditsche-Kuru P, Gallenmüller M, Gude M, Haushahn T, Hermann M, Immink H, Knippers J, Lienhard J, Luchsinger R, Lunz K, Mattheck C, Milwich M, Mölders N, Neinhuis C, Nellesen A, Poppinga S, Rechberger M, Schleicher S, Schmitt C, Schwager H, Seidel R, Speck O, Stegmaier T, Tesari I, Thielen M, Speck T (2012) Biomimetic products. In: Bar-Cohen Y (ed) Biomimetics: nature-based innovation. CRC Press / Taylor & Francis Group, Boca Raton, London, New York, pp 377–429

8. Speck T, Speck O (2008) Process sequences in biomimetic research. In: Brebbia CA (ed) Design and nature IV. WIT Press, Southampton, pp 3–11

9. Speck, O, Luchsinger R, Rampf M and Speck T (2014) Selbstreparatur in Natur und Technik. Konstruktion 9:72–75 + 82

10. Busch S, Seidel R, Speck O, Speck T (2010) Morphological aspects of self-repair of lesions caused by internal growth stresses in stems of *Aristolochia macrophylla* and *Aristolochia ringens*. Proc R Soc Lond. B 277:2113–2120

11. Rampf M, Speck O, Speck T, Luchsinger R (2013) Investigation of a fast mechanical self-repair mechanism for inflatable structures. Int J Eng Sci 63:61–70

12. Bauer G, Speck T (2012) Restoration of tensile strength in bark samples of *Ficus benjamina* due to coagulation of latex during fast self-healing of fissures. Ann Bot 109:807–811

13. Bauer G, Friedrich C, Gillig C, Vollrath F, Speck T, Holland C (2014) Investigating the rheological properties of native plant latex. J R Soc Inter 11(90). https://doi.org/10.1098/rsif.2013.0847

14. Bauer G, Gorb S, Klein MC, Nellesen AV, Tapavicza M, Speck T (2014) Comparative study on latex particles and latex coagulation in *Ficus benjamina*, *Campanula glomerata* and three *Euphorbia* species. PLoS ONE 9(11):e113336. https://doi.org/10.1371/journal.pone.0113336

15. Schüssele AC, Nübling F, Thomann Y, Carstensen O, Bauer G, Speck T, Mülhaupt R (2012) Self-healing rubbers based on NBR blends with hyperbranched polyethylenimines. Macromol Mater Eng 297:411–419

16. Thielen M, Schmitt CNZ, Eckert S, Speck T, Seidel R (2013) Structure-function relationship of the foam-like pomelo peel (*Citrus maxima*)—an inspiration for the development of biomimetic damping materials with high energy dissipation. Bioinspir. Biomim. 8. https://doi.org/10.1088/1748-3182/8/2/025001

17. Fischer SF, Thielen M, Loprang RR, Seidel R, Fleck C, Speck T, Bührig-Polaczek A (2010) Pummelos as concept generators for biomimetically-inspired low weight structures with excellent damping properties. Adv Eng Mater/Adv Biomater 12:B658–B663

18. Thielen M, Speck T, Seidel R (2013) Viscoelasticity and compaction behaviour of the foam-like pomelo (*Citrus maxima*) peel. J Mater Sci 48:3469–3478

19. Thielen M, Speck T, Seidel R (2015) Impact behaviour of freeze-dried and fresh pomelo (*Citrus maxima*) peel - influence of the hydration state. R Soc Open Sci 2:140322. https://doi.org/10.1098/rsos

20. Fischer SF, Thielen M, Weiß P, Seidel R, Speck T, Bührig-Polaczek A, Bünck M (2014) Production and properties of a precision-cast bio-inspired composite. J Mater Sci 49:43–51

21. Spatz HC, Beismann H, Brüchert F, Emanns A, Speck T (1997) Biomechanics of the giant reed *Arundo donax*. Philos Trans R Soc Lond B 352:1–10
22. Speck T, Speck O, Emanns A, Spatz HC (1998) Biomechanics and functional anatomy of hollow stemmed sphenopsids: III. *Equisetum hyemale*. Botanica Acta 111:366–376

Mechanics of Ageing—From Building to Biological Materials

B. A. Schrefler, F. Pesavento, R. Santagiuliana and G. Sciumè

Abstract In this work, we present a general model for the analysis of concrete and biological tissues as multiphase porous materials, with particular regard to their ageing. Such problems are typically multiphysics ones with overlapping domains where diffusion, advection, adsorption, phase changes, deformation, chemical reactions and other phenomena take place in the porous medium. For the analysis of such a complex system, the model here proposed is obtained from the microscopic scale by applying the Thermodynamically Constrained Averaging Theory (TCAT) which guarantees the satisfaction of the second law of thermodynamics for all constituents both at micro- and macrolevels. Moreover, one can obtain some important thermodynamic restrictions imposed on the evolution equations describing the material deterioration. Two specific forms of the general model adapted to the cases of cementitious and biological materials respectively are shown. Some numerical simulations, aimed at proving the validity of the approach adopted, are also presented and discussed.

1 Introduction

Most materials show an evolution of their characteristics over time, or with use (for better or worse). This change of their properties often corresponds to a degradation of their performances from physical, chemical and mechanical points of view due to a natural process, to more-or-less intensive use and to the action of external agents (biological, chemical or physical). Corrosion, obsolescence, and weathering are examples of ageing.

B. A. Schrefler · F. Pesavento (✉) · R. Santagiuliana
Department of Civil, Environmental and Architectural Engineering,
University of Padova, Via F. Marzolo 9, 35131 Padova, Italy
e-mail: francesco.pesavento@dicea.unipd.it

G. Sciumè
Institut de Mécanique et d'Ingénierie I2M – Dép. TREFLE, Université de Bordeaux,
ENSAM - Esplanade des Arts et Métiers, 33405 Talence Cedex, France

© Springer International Publishing AG 2018
K. van Breugel et al. (eds.), *The Ageing of Materials and Structures*,
https://doi.org/10.1007/978-3-319-70194-3_4

In this work, our attention is focused on the ageing of two classes of materials: cementitious materials and biological materials involved in the tumour formation and development.

The common feature of these two types of material is that both can be considered as porous in nature. The voids in their microstructure can be filled with fluids, of liquid or gaseous form. Several chemical species can be dispersed in the fluid phases, and strong interactions between these chemical species, different phases/components and the solid skeleton can be observed.

These interactions range from the physical ones (e.g. the phase changes), to chemical ones (e.g. chemical reaction taking places between the constituents under certain conditions) and/or to mechanical ones (for instance, the pressures exerted by the fluid phases on the solid matrix).

For the analysis of such a complex systems, models based on standard formulations are not enough for capturing the behaviour of the material in a realistic manner. It is necessary to consider Porous Media Mechanics.

In the last couple of decades, we have developed two major three-fluid flow models for deforming porous solids. The first is our three-fluids model for concrete which has been in development since the late 1990s and extended to many applications in concrete technology, such as concrete under high temperatures [6–8, 11], leaching in isothermal and non-isothermal conditions [13–15], Alkali-Aggregate Reactions [22], young concrete and repair problems [9, 10, 21, 26] and recently also freezing/thawing [20]. It considers concrete as formed by a deformable solid matrix, the pores of which are filled with dry air, water vapour and liquid water. The interactions between all the constituents are duly taken into account. The second one is our recently developed tumour-growth model which considers tumours as comprising various compartments: the extracellular matrix as deformable porous solid, the pores of which are filled with healthy cells, tumour cells, both living and necrotic and the interstitial fluid which carries nutrients, therapeutic agents, and waste products [1, 25, 27–31]. The vasculature/neovasculature is a separate compartment with only mass exchange with the other compartments. The cells are treated as adhesive fluids. Both models are hence de facto three-fluid flow models for a deforming porous medium. The same set of material free-balance equations apply to these apparently completely different problems, governing the structure of the code for numerical solution. This insight enabled us to build the second model in very short time, starting from the concrete model even though the constitutive behaviour and evolution laws differ substantially. Both models are implemented in the code CAST3M (http://www-cast3m.cea.fr) of the French Atomic Energy Commission (CEA) and are easily accessible. In the following, we show the balance equations for both cases to seek to convince the reader of the analogy, as well as explaining a recent application for each of the two cases.

Some numerical examples are presented for ageing concrete and growing tumour masses. The results will highlight the suitability of the proposed multiphase models for the prediction of evolution/aging of biological and cementious materials. Perspectives for future developments are outlined in the conclusions section.

2 Three-Fluids Model for Concrete and Sample Applications

The balance equations are written by considering concrete as a multiphase porous material assumed to be in thermal-, hygral- and mechanical-equilibrium state locally, even though chemical reactions (a hydration process) are progressing at certain rate. The local chemical equilibrium is reached only at the end of the spontaneous hydration process and, from that time, a local thermodynamic equilibrium state is reached at a given point. Characteristic times of the local thermal, hygral and mechanical processes are much smaller than that of the hydration reaction; hence, for these processes we use the equilibrium relations (with the actual value of hydration degree being 'frozen') [9]. The solid skeleton voids are filled partly by liquid water and partly by a gas phase. The liquid phase consists of bound water, which is present in the whole range of moisture content, and capillary water, which appears when the degree of water saturation exceeds the upper limit of the hygroscopic region. The gas phase is a mixture of dry air and water vapour (condensable gas constituent) and is assumed to behave as an ideal gas. The model equations are obtained by means of the Thermodynamically Constrained Averaging Theory TCAT [16, 17] which is used to transform known microscale relations to mathematically and physically consistent macroscale relations. Here the balance equations are written already at macroscopic level for the sake of brevity.

The *mass balance equation for the solid phase s* is

$$\frac{\partial(\varepsilon^s \rho^s)}{\partial t} + \nabla \cdot \left(\varepsilon^s \rho^s \bar{\mathbf{v}}^s\right) = \overset{l \to Hs}{M} \tag{1}$$

The *mass balance equation of liquid water l* is

$$\frac{\partial(\varepsilon^l \rho^l)}{\partial t} + \nabla \cdot \left(\varepsilon^l \rho^l \bar{\mathbf{v}}^l\right) = -\overset{l \to Hs}{M} - \overset{l \to Wg}{M} \tag{2}$$

The *mass balance of water vapour Wg* reads

$$\frac{\partial\left(\varepsilon^g \rho^g \omega^{\overline{Wg}}\right)}{\partial t} + \nabla \cdot \left(\varepsilon^g \rho^g \omega^{\overline{Wg}} \bar{\mathbf{v}}^{\bar{g}}\right) + \nabla \cdot \left(\varepsilon^g \rho^g \omega^{\overline{Wg}} \mathbf{u}^{\overline{\overline{Wg}}}\right) = \overset{l \to Wg}{M} \tag{3}$$

and that *of dry air Ag* is

$$\frac{\partial\left(\varepsilon^g \rho^g \omega^{\overline{Ag}}\right)}{\partial t} + \nabla \cdot \left(\varepsilon^g \rho^g \omega^{\overline{Ag}} \bar{\mathbf{v}}^{\bar{g}}\right) + \nabla \cdot \left(\varepsilon^g \rho^g \omega^{\overline{Ag}} \mathbf{u}^{\overline{\overline{Ag}}}\right) = 0 \tag{4}$$

where $\overset{l \to Hs}{M}$ is the chemically combined water, $\overset{l \to Wg}{M}$ the vapourized water, g indicates the gaseous phase, mixture of dry air and water, ε the volume fraction, ω the mass fraction, $\mathbf{v}$ the phase velocity, $\mathbf{u}$ the relative velocity and ρ the density.

Finally, the *linear momentum balance equation* is

$$\nabla \cdot \left(\frac{\partial \mathbf{t}^s_{eff}}{\partial t} - \frac{\partial p^s}{\partial t} \mathbf{1} \right) = 0 \tag{5}$$

where $\mathbf{t}^s_{eff}$ is the effective stress and p^s the solid pressure [12, 18, 19]. These balance equations have to be supplemented with appropriate constitutive equations and evolution laws for which the reader is referred to the relevant papers in the references. We limit ourselves here to a brief description of the evolution model for the analysis of the hydration process in concrete at early ages.

2.1 Hydration Model for Ageing Concrete

For the analysis of the behaviour of cementitious materials at early ages, as, for example, concrete during the hydration process, or the long-term behaviour of concrete structures, a model describing the development of the chemical reactions involved is needed.

So, in addition to what was described in the previous sections, it is necessary to formulate an evolution equation accounting for the chemical reactions that the hydration/hardening process is based on. To meet this requirement, a non-dimensional measure related to the chemical reaction extent, known as hydration degree, can be defined as follows:

$$\Gamma_{hydr} = \frac{\chi}{\chi_\infty} = \frac{m_{hydr}}{m_{hydr\infty}} \tag{6}$$

where m_{hydr} means mass of hydrated water (chemically combined), χ is the hydration extent and χ, m_{hydr} are the final values of hydration extent and mass of hydrated water, respectively.

From the macroscopic point of view, hydration of cement is a complex interactive system of competing chemical reactions of various kinetics and amplitudes. They are associated with complex physical and chemical phenomena at the microlevel of material structure, resulting in considerable changes of macroscopic concrete properties. Kinetics of cement hydration (hydration rate) cannot be described properly in terms of an equivalent age nor maturity of concrete, if the effect of the reaction rate on temperature (and/or relative humidity) depends upon the hydration degree, or chemical affinity of the reaction is affected by temperature variations (and/or relative humidity) [9]. Hence, another thermodynamically based approach has been used instead, similar to that proposed by Ulm and Coussy [32], see [9, 10]. In this approach, the hydration extent χ is the advancement of the hydration reaction, and its rate is related to the affinity of the chemical reaction through an Arrhenius-type relationship, as is usual for thermally activated chemical reactions [9, 10]:

$$\frac{d\chi}{dt} = \tilde{A}_\chi(\chi) \exp\left(-\frac{E_a}{RT}\right) \tag{7}$$

where $\tilde{A}_\chi$ is the normalized affinity (it accounts both for chemical non-equilibrium and for the nonlinear diffusion process), E_a is the hydration activation energy and R is the universal gas constant. Equation (7) can be rewritten in terms of hydration degree, defined as in Eq. (6), and relative humidity by means of a function $\beta_\varphi(\varphi)$, (φ is the relative humidity) [9, 10]:

$$\frac{d\Gamma_{hydr}}{dt} = \tilde{A}_\Gamma(\Gamma_{hydr})\beta_\varphi(\varphi) \exp\left(-\frac{E_a}{RT}\right) \tag{8}$$

An analytical formula for the description of the normalized affinity of the following form:

$$\tilde{A}_\Gamma(\Gamma_{hydr}) = A_1\left(\frac{A_2}{\kappa_\infty} + \kappa_\infty \Gamma_{hydr}\right)(1 - \Gamma_{hydr}) \exp(-\bar{\eta}\,\Gamma_{hydr}) \tag{9}$$

was proposed by Cerver et al. [3] and is used in our model. The coefficients A_1, A_2 and $\bar{\eta}$ can be obtained from the temperature evolution during adiabatic tests.

Taking into account that the material properties change with time, all the properties used in the model are a function of hydration degree.

2.2   Modeling of a Repaired Beam

The numerical simulation of the thermo-hygro-mechanical behaviour of two repaired beams is presented in this subsection (see [26]. Numerical results are compared with the experimental ones of the reference experiment performed by Bastien Masse [2].

The geometry of the reinforced beams is represented in Fig. 1. Three identical reinforced beams were cast for the experiment. At thirty days after the casting, two

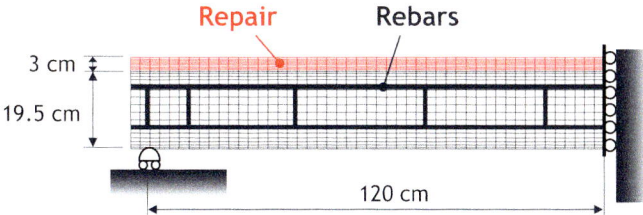

Fig. 1 Geometry and FE mesh of the repaired beam. (Reproduced from Sciumè et al. [26])

of these beams, after the hydro-demolition of 30 mm of the upper part, had been repaired: one using the ordinary concrete (OC) (very similar to that used to cast the three beams), and the other using ultra-high-performance fibre-reinforced concrete (UHPC). The third beam is the reference specimen. Two fibre-optic sensors (FO-h and FO-b) were placed inside the beams. More details on the experimental procedure (laboratory conditions, specimens equipment etc.) can be found in Bastien Masse [2].

The beams have been modelled in 2D-plane stress; for more information about the assumed boundary conditions and the FE discretization, refer to [26].

It is important to highlight that the THCM history of the two beams before the repair and the wetting procedure for the preparation of the substrate have been also taken into account within the modeling process (the numerical simulations start from the casting of the beams), and this has been critical to succeed in predictive numerical results.

Before performing the simulation of the beams the input parameters for the two considered concretes have been identified. This has been done exploiting experiments performed by Bastien Masse [2] for the characterization of the two repair materials. Comparison between experimental results and numerical ones for adiabatic calorimetry, creep properties, Young's modulus and Poisson's ration evolution and autogenous and drying shrinkage are shown in Fig. 2.

In Fig. 3, for the two repair case, the relative humidity at one hour, one week and one month after repair is depicted. In the UHPC, self-desiccation has also a relevant impact on the decrease of relative humidity.

Figure 4 shows the numerical and the experimental results for the vertical displacement of the three beams measured using the linear potentiometer placed in the lower-middle point of the beams. A good agreement between the experimental results and the numerical ones can be observed. The deflection of the reference beam is mainly due to its weight and also to the non-symmetric position of the steel reinforcements: in other words, the shrinkage of the upper and lower parts of the beam generates an eccentric force that increases the deflection of the beam. In the repaired beams, the deflections are accentuated by the autogenous and drying shrinkage of the fresh restoration materials.

After two months, the two repaired beams have been submitted to a three-point bending test, and also, in this case, the presented model has shown good accuracy in predicting experimental results [26].

3   Three-Fluids Model for Tumour Growth and Applications

From review papers of numerical models for tumour growth, such as Roose et al. [23], Deisboeck et al. [5] and Sciumè et al. [28], it appears clear that, in the realm of a continuum approach, a vast majority of models describe the malignant mass (TC), the host cells (HC) and the interstitial fluid (IF) as homogeneous, viscous fluids and

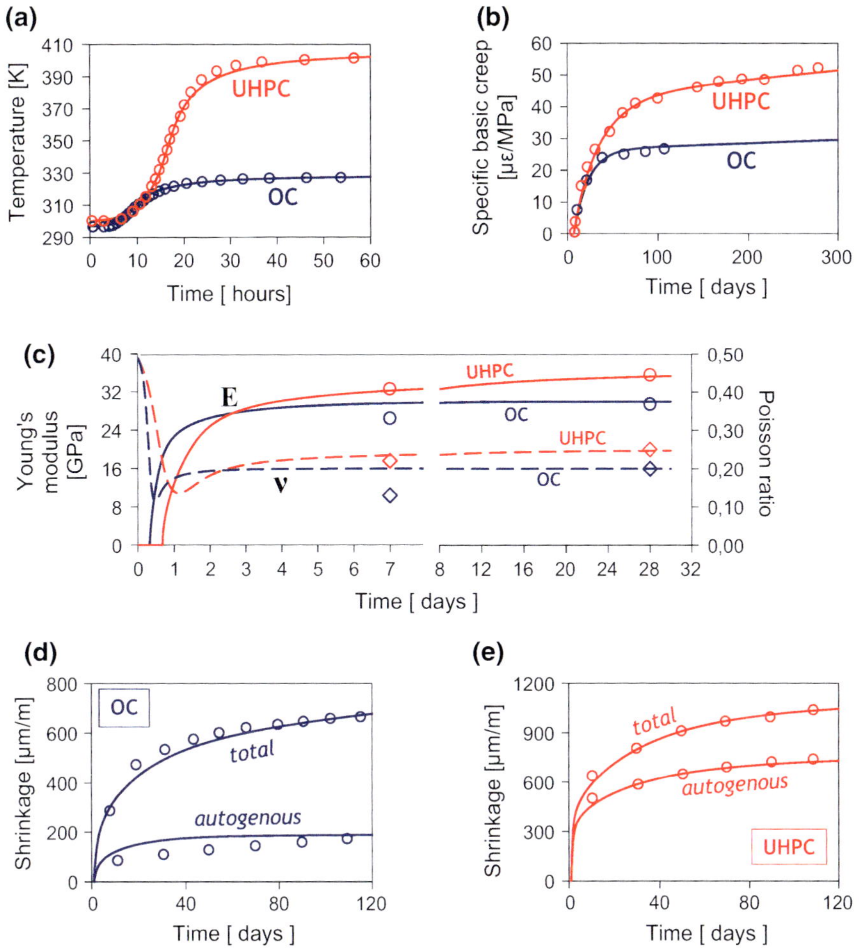

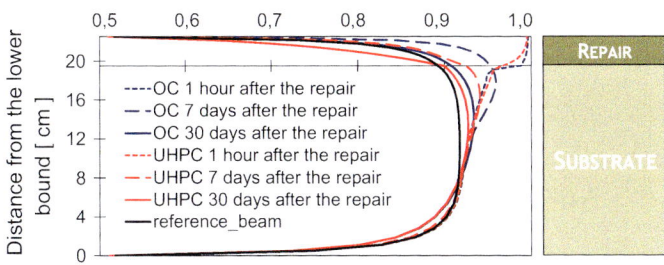

Fig. 2 Experimental tests exploited to identify the model parameters: **a** adiabatic calorimetry, **b** creep test, **c** Young's modulus and Poisson's ration evolution, **d** autogenous and drying shrinkage for OC, **e** autogenous and drying shrinkage for UHPC. Symbols are experimental data while solid and dashed lines are numerical results. Rearrangement of figures from Sciumè et al. [26]

Fig. 3 Relative humidity at one hour, one week and one month after repair for the two simulated cases [26]

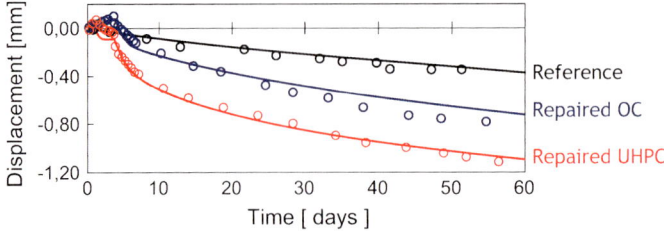

Fig. 4 Experimental (open symbols) and numerical results (solid lines) for the vertical displacement of the middle points of the three beams. The time "zero" corresponds to the application of the repairs. Reprinted from Sciumè et al. [26]

employ reaction-diffusion-advection equations for predicting the distribution and transport of nutrients. The Extracellular matrix (ECM) is absent in almost all the models.

In our model, the tumour comprises the following phases: (i) tumour cells (TC), which partition into living cells (LTC) and necrotic cells (NTC); (ii) healthy cells (HC); (iii) extracellular matrix (ECM); and (iv) interstitial fluid (IF); see Sciumè et al. [25, 28–31]. The ECM is a porous solid, while all other phases are fluids. The tumour cells may become necrotic upon exposure to low-nutrient concentrations or excessive mechanical pressure. The IF is a mixture of water and biomolecules, such as nutrients, oxygen, therapeutic agents and waste products. In the following mass and momentum balance equations, α denotes a generic phase, t the tumor cells (TC), h the healthy cells (HC), s the solid phase (ECM) and l the interstitial fluid (IF).

The balance equations are listed below to show the analogy with the concrete model. For the full development the reader is referred to Sciumè et al. [27, 29–31].

The *mass balance equation of the solid s* is

$$\frac{\partial(1-\varepsilon)}{\partial t} + \nabla \cdot \left[(1-\varepsilon)\mathbf{v}^s\right] = 0 \tag{10}$$

where $\mathbf{v}^s$ is the solid phase velocity.

The mass balance equations of the tumour cell phase t, the host cell phase h and interstitial fluid l are respectively:

$$\frac{\partial(\varepsilon S^t)}{\partial t} + \nabla \cdot \left(\varepsilon S^t \mathbf{v}^s\right) - \nabla \cdot \left(\frac{k_{rel}^t \bar{\mathbf{k}}^{ts}}{\mu^t} \langle 1 - \frac{a_t}{|\nabla p^t|} \rangle_+ \nabla p^t \right) = \frac{1}{\rho} \overset{l \to t}{M}_{growth} \tag{11}$$

$$\frac{\partial(\varepsilon S^h)}{\partial t} + \nabla \cdot \left(\varepsilon S^h \mathbf{v}^s\right) - \nabla \cdot \left(\frac{k_{rel}^h \bar{\mathbf{k}}^{hs}}{\mu^h} \langle 1 - \frac{a_h}{|\nabla p^h|} \rangle_+ \nabla p^h \right) = 0 \tag{12}$$

$$\frac{\partial(\varepsilon S^l)}{\partial t} + \nabla \cdot \left(\varepsilon S^l \mathbf{v}^s\right) - \nabla \cdot \left(\frac{k_{rel}^l \bar{\mathbf{k}}^{ls}}{\mu^l} \nabla p^l \right) = -\frac{1}{\rho} \overset{l \to t}{M}_{growth} \tag{13}$$

where a_α is the adhesion ($al = 0$), μ^α is the dynamic viscosity, k_{rel}^α is the relative permeability which takes care of the presence of the other two fluid phases [29], $\bar{\mathbf{k}}^{as}$ is the absolute permeability, p^α is the pressure and ρ is the common density; $\underset{growth}{\overset{l \rightarrow t}{M}}$ is the rate of growth term. The linear-momentum balance equations of the fluid phases have here already been introduced to show the difference between cells and IF.

Assuming that: (i) there is no diffusion of either necrotic or living cells; (ii) there is no exchange of necrotic cells with other phases, the *mass conservation equation for the necrotic portion of the tumour cells phase* reads:

$$\frac{\partial \left(\varepsilon^t \rho \omega^{N\bar{t}} \right)}{\partial t} + \nabla \cdot \left(\varepsilon^t \rho \omega^{N\bar{t}} \mathbf{v}^{\bar{t}} \right) - \varepsilon^t r^{Nt} = 0 \tag{14}$$

where $\omega^{N\bar{t}} = \frac{\varepsilon^{N\bar{t}} \rho^{N\bar{t}}}{\varepsilon^t \rho^t}$ is the mass fraction of necrotic cells in the tumour cells phase t, $\varepsilon^t r^{Nt}$ the rate of generation of necrotic cells and $\mathbf{v}^{\bar{t}}$ the velocity of the tumour cells phase.

The *mass-balance equation of the nutrient species* in the interstitial fluid is

$$\frac{\partial \left(\varepsilon S^l \omega^{\overline{nl}} \right)}{\partial t} + \nabla \cdot \left(\varepsilon S^l \omega^{\overline{nl}} \mathbf{v}^{\bar{l}} \right) - \nabla \cdot \left(\varepsilon S^l D_{eff}^{\overline{nl}} \nabla \omega^{\overline{nl}} \right) = - \frac{\overset{nl \rightarrow t}{M}}{\rho^l} \tag{15}$$

where $D_{eff}^{\overline{nl}}$ is the effective diffusion coefficient depending on the available pore space [29] and $\overset{nl \rightarrow t}{M}$ is the nutrient consumption rate which depends on the local nutrient availability. Finally, the *linear-momentum balance equation* of the solid phase in rate form reads:

$$\nabla \cdot \left(\frac{\partial \mathbf{t}_{eff}^s}{\partial t} - \frac{\partial p^s}{\partial t} \mathbf{1} \right) = 0 \tag{16}$$

where $\mathbf{t}_{eff}^s$ is the effective stress tensor in the solid and $\mathbf{1}$ is the unit tensor. The interaction between the solid and the fluids, inclusive of the cell populations, has been accounted for through the effective stress principle [18, 19]. Clearly, these balance equations have to be completed with the appropriate constitutive equations, which can be found in Sciumè et al. [27, 29–31].

3.1 Applications of the Tumour-Growth Model

In 2011, when we started working on the tumour-growth model, we introduced two simplifying assumptions: (i) a unique pressure was considered for both cell populations ($p^t = p^h$) and (ii) the ECM was assumed rigid (see [27]).

Then, the introduction of the relevant constitutive relationships for the pressure difference between each pair of fluid phases has allowed for relaxation of the first hypothesis and for a more realistic modeling of cell adhesion and invasion; this is shown in [29].

In Sciumè et al. [31], the second hypothesis also has been relaxed and nowadays ECM deformability and its impact on tumour growth can be properly taken into account, as shown in the second presented example, which deals with growth of a melanoma within a deformable ECM. The first example shows a comparison between experimental and numerical results with regard to the MTS growth in vitro.

3.1.1 MTS Growth in Vitro: Comparison with Experimental Data

With the enhanced version of the model presented in Santagiuliana et al. [24], we reproduce the available data of Multicellular Tumour Spheroid (MTS) growth experiments in vitro carried out at Houston Methodist Research Institute HMRI. Differently from [31], the simulation is performed with free boundary conditions; this means that we simulate only the MTS, without IF or ECM outside the border, see Fig. 5. The rim of MTS moves during the time with velocity:

$$\mathbf{v}^t = -\frac{k_{rel}^t \mathbf{k}}{\mu^t \varepsilon_t} \cdot \nabla p^t \tag{17}$$

Hence the boundary conditions are updated at every step and have fixed values on the moving boundary: zero IF pressure and oxygen concentration equal to 7.0×10^{-6} Pa.

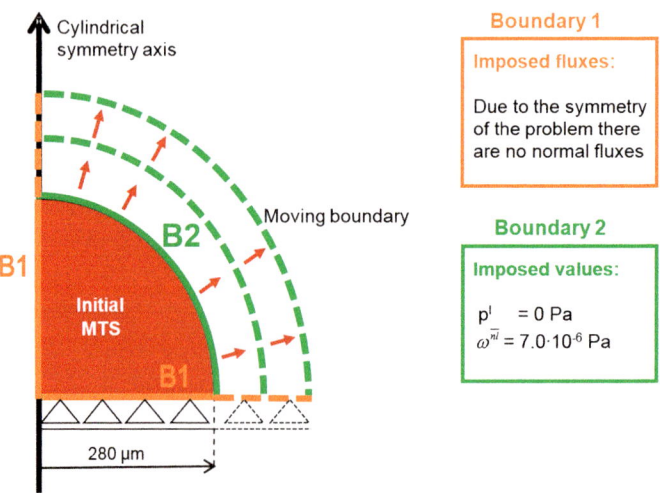

Fig. 5 Free boundary conditions. Geometry and boundary conditions for an MTS growing with moving boundary [24]

Table 1 Initial conditions

Zone	p^l(Pa)	S^h (−)	S^t(−)	$\overline{\omega}^{nl}$(−)
Red zone (up to 280 μm)	0.0	0.0	0.01	7.0×10^{-6}

Fig. 6 Model results and experimental data. Comparison between model results (solid line) and experimental data (dots) U87 spheroid cultivated by C. Stigliano at HMRI. The red line refers to the case of the MTS growing in an ECM deposited by TCs and the blue line to the case of the MTS growing in an ECM-free culture medium

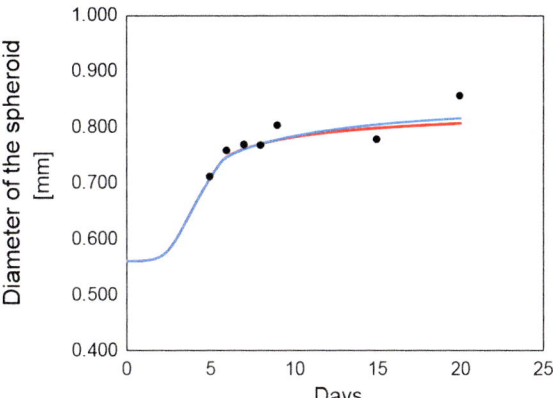

The initial conditions in the MTS zone are: volume fraction of TCs equal to 0.01, volume fraction of HC equal to 0 and porosity equal to 1 (see Table 1). A deformable ECM is assumed, and it is deposed by TCs during the tumour-mass growth. For the parameters of the cells for this simulation, please see [24], where also the culture of these MTS made of human multiforme glioblastoma U-87, MG cells (ATCC) are described in detail.

The results are shown in Fig. 6, where dots refer to the diameter of U87 spheroid. The simulations capture both the first exponential phase of the growth and the second phase: the plateau where the tumour mass reaches an equilibrium. A steady radius, as documented in several experiments, characterises the final phase of spheroid growth. The equilibrium, in the simulations, is reached thanks to the free boundary conditions and to the growth limiter due to pressure. The moving boundary with a Dirichlet condition for the nutrient concentration at the MTS surface allows simulating growth without such a condition posed at some point in the culture medium that may influence the proliferation and the movement of tumour cells. The red line in Fig. 3 refers to the case of the MTS growing in an ECM deposited by TCs, while the blue line refers to the case of the MTS growing in an ECM-free culture medium. In this second case, the steady diameter in the final phase of the MTS growth is slightly larger than in the first case. The explanation could be that the stiffness of the ECM hinders the increase of TCs. Hence, the viscoplastic law (see [31]) used for the solid phase behaviour seems to influence the MTS growth. Without ECM, instead, the tumour cells growth is not confined.

3.1.2 Growth of a Melanoma

The second case deals with growth of a skin melanoma during the avascular stage.

Cutaneous melanoma is the most dangerous form of skin cancer. It arises in the melanocytes, the specialized melanin producing cells, which are scattered along the epidermis-dermis border, see Fig. 7. It is recalled that the outer structure of the skin has a layered structure where three compartments can be evidenced: the epidermis, an outer epithelium of stratified cells; the dermis, an intermediate cushion of vascularized connective tissue; and the hypodermis, the lowermost layer made of loose tissue and adipose cells. The dermis is separated from the epidermis by the basement membrane or basal laminae, a tough sheet of ECM, see Fig. 7.

Consistently with the structure of the skin, we consider three zones: (i) an epidermal zone of 185 μm (EZ); (ii) a basement membrane of 30 μm (MZ) and a (iii) dermal zone of 585 μm (DZ) (see Fig. 7). A different volume fraction of ECM is assumed for each of these three zones according to their different natures ($\varepsilon_{EZ}^s = 0.2$ for the EZ, $\varepsilon_{MZ}^s = 0.3$ for the MZ and $\varepsilon_{DZ}^s = 0.1$ for the DZ).

EZ, MZ and DZ having their own porous microstructure, a Young's modulus, an intrinsic permeability and a p-s constitutive relationship have been specified for each zone. Because of the different p-s constitutive relationships, discontinuities of the saturation degrees of fluid phases are allowed across interfaces EZ-MZ and MZ-DZ, while the pressure fields are continuous. This is an advantage of the adopted pressure-based formulation. The problem of Fig. 7 has been modelled exploiting cylindrical symmetry.

For details on the input parameters, initial and boundary conditions, refer to [31].

Figure 8 shows the melanoma after two weeks of growth and a qualitative comparison with a clinical observed case of [4].

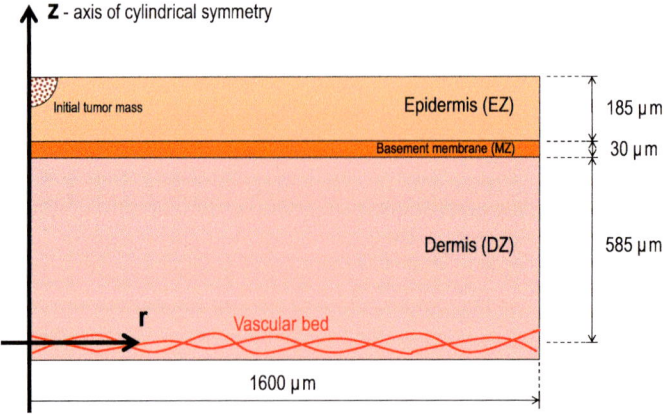

Fig. 7 Skin structure and geometry of the modelled case. (Sciumè et al. [31])

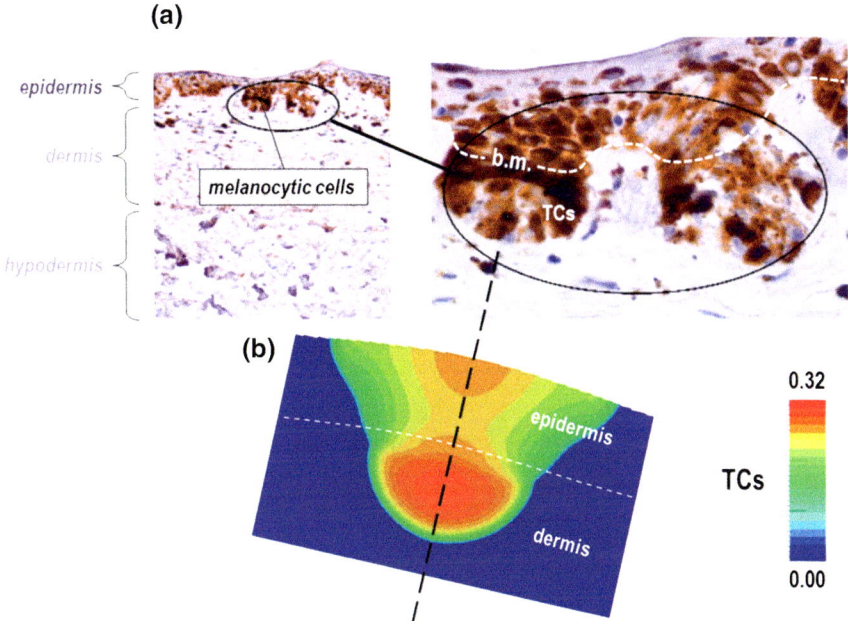

Fig. 8 Paraffin-embedded tissue sections of a melanocytic lesion, [4]; b.m. stands for basement membrane (**a**); numerical results (volume fraction of TCs) after two weeks of growth (**b**) [31]

4 Conclusions and Perspectives

In this work, we have presented a general approach based on Multiphase Porous Media Mechanics for the analysis of the behaviour of cementitious and biological materials, with particular regard to ageing. In this context, ageing is interpreted as any change of the properties of the material due to natural processes, to more-or-less intensive use and to the action of external agents (biological, chemical or physical).

The two classes of materials taken into consideration here can be seen as multiphase porous materials because several fluids fill the solid-matrix pore network, with various interactions between the phases and the components. Thus, the approach based on MPMM seems the most suitable one for the analysis of such materials.

The model proposed is obtained by applying the Thermodynamically Constrained Averaging Theory (TCAT) which guarantees the upscaling of the entire thermodynamics from the microlevel. The set of governing equations at macroscopic level has been shown and described.

The general model has been particularised for the case of concrete ("concrete model") and biological material ("tumour model"), taking into account specific form of the equations, the constitutive relationships and the evolution laws. Some applications of the model to concrete structures (modelling of a repaired beam) and to the simulation of tumour development (MTS growth and melanoma growth) have been presented and discussed.

These applications show the validity and flexibility of the general approach adopted in this work, which allows for the simulation of the behaviour of materials and multiphase systems that are apparently very different in nature.

Acknowledgements The authors thank Cinzia Stigliano of HMRI for the provided experimental data. RS acknowledges the University of Padua for financial support (project n. CPDR121149). FP acknowledges the University of Padua for financial support (project n. CPDA135049).

References

1. Bao G, Bazilevs Y, Chung J-H, Decuzzi P, Espinosa HD, Ferrari M, Gao H, Hossain SS, Hughes TJR, Kamm RD, Liu WK, Marsden A, Schrefler B (2014) USNCTAM perspectives on mechanics in medicine. J Royal Soc Interface 11:20140301
2. Bastien Masse, M (2010) Étude du comportement déformationnel des bétons de réparation. Master Thesis. Université de Montréal
3. Cervera M, Olivier J, Prato T (1999) A thermo-chemo-mechanical model for concrete. II: damage and creep. J Eng Mech (ASCE) 125(9):1028–1039
4. Chung LS, Man Y-G, Lupton GP (2010) WT-1 expression in a spectrum of melanocytic lesions: Implication for differential diagnosis. J Cancer 1:120
5. Deisboeck TS, Wang Z, Macklin P, Cristini V (2011) Multiscale cancer modeling. Annu Rev Biomed Eng 13:127–155
6. Gawin D, Pesavento F, Schrefler BA (2002) Modelling of hygro-thermal behaviour and damage of concrete at temperature above the critical point of water. Int J Numer Anal Methods Geomech 26(6):537–562
7. Gawin D, Pesavento F, Schrefler BA (2003) Modelling of hygro-thermal behaviour of concrete at high temperature with thermo-chemical and mechanical material degradation. Comput Methods Appl Mech Eng (CMAME) 192(13–14):1731–1771
8. Gawin D, Pesavento F, Schrefler BA (2004) Modelling of deformations of high strength concrete at elevated temperatures. Mater Struct 37(268):218–236
9. Gawin D, Pesavento F, Schrefler BA (2006) Hygro-thermo-chemo-mechanical modelling of concrete at early ages and beyond. Part I: Hydration and hygro-thermal phenomena. Int J Numer Meth Eng 67(3):299–331
10. Gawin D, Pesavento F, Schrefler BA (2006) Hygro-thermo-chemo-mechanical modelling of concrete at early ages and beyond. Part II: shrinkage and creep of concrete. Int J Numer Methods Eng 67(3):332–363
11. Gawin D, Pesavento F, Schrefler BA (2006) Towards prediction of the thermal spalling risk through a multi-phase porous media model of concrete. Comput Methods Appl Mech Eng 195:5707–5729
12. Gawin D, Pesavento F, Schrefler BA (2007) Modelling creep and shrinkage of concrete by means of effective stress. Mater Struct 40:579–591
13. Gawin D, Pesavento F, Schrefler BA (2008) Modeling of cementitious materials exposed to isothermal calcium leaching, with considering process kinetics and advective water flow. Part 1: theoretical model. Solids Struct 45:6221–6240

14. Gawin D, Pesavento F, Schrefler BA (2008) Modeling of cementitious materials exposed to isothermal calcium leaching, with considering process kinetics and advective water flow. Part 2: numerical solution. Solids Struct 45:6241–6286

15. Gawin D, Pesavento F, Schrefler BA (2009) Modeling deterioration of cementitious materials exposed to calcium leaching in non-isothermal conditions. Comput Methods Appl Mech Eng 198:3051–3083. https://doi.org/10.1016/j.cma.2009.05.005

16. Gray WG, Miller CT, Schrefler BA (2013) Averaging theory for description of environmental problems: what have we learned? Adv Water Res 51:123–138

17. Gray WG, Miller CT (2014) Introduction to the thermodynamically constrained averaging theory for porous medium systems. Springer, New York

18. Gray WG, Schrefler BA (2001) Thermodynamic approach to effective stress in partially saturated porous media. Eur J Mech A/Solids 20:521–538

19. Gray WG, Schrefler BA, Pesavento F (2009) The solid phase stress tensor in porous media mechanics and the Hill-Mandel condition. J Mech Phys Solids 57:539–554

20. Koniorczyk M, Gawin D, Schrefler BA (2015) Modelling evolution of frost damage in fully saturated porous materials. Comput Methods Appl Mech Eng 297:38–61

21. Pesavento F, Gawin D, Schrefler BA (2008) Modeling cementitious materials as multiphase porous media: theoretical framework and applications. Acta Mech 201:313–339. https://doi.org/10.1007/s00707-008-0065-z

22. Pesavento F, Gawin, D, Wyrzykowski, M, Schrefler, BA and Simoni, L (2012) Modeling alkali-silica reaction in non-isothermal, partially saturated cement based materials. Comput Methods Appl Mech Eng (CMAME), 225–228:95–115. https://doi.org/10.1016/j.cma.2012.02.019

23. Roose T, Chapman SJ, Maini PK (2007) Mathematical models of avascular tumor growth. Siam Rev 49(2):179–208

24. Santagiuliana R, Stigliano C, Mascheroni P, Ferrari M, Decuzzi, P, Schrefler BA (2015) The role of cell lysis and matrix deposition in tumor growth modeling. Adv Model Simul Eng Sci 2:19. https://doi.org/10.1186/s40323-015-0040-x455

25. Sciumè G, Shelton SE, Gray WG, Miller CT, Hussain F, Ferrari M, Decuzzi P, Schrefler BA (2012) Tumor growth modeling from the perspective of multiphase porous media mechanics. Mol Cell Biomech 9(3):193–212

26. Sciumè G, Benboudjema F, De Sa C, Pesavento F, Berthaud Y, Schrefler BA (2013) A multiphysics model for concrete at early age applied to repairs problems. Eng Struct 57:374–387

27. Sciumè G, Shelton S, Gray WG, Miller CT, Hussain F, Ferrari M, Decuzzi, P, Schrefler BA (2013) A multiphase model for three-dimensional tumor growth. New J Phys 15:015005

28. Sciumè G, Gray WG, Ferrari M, Decuzzi P, Schrefler BA (2013) On computational modeling in tumor growth. Arch Comput Methods Eng 20:327–352

29. Sciumè G, Gray WG, Hussain F, Ferrari M, Decuzzi P, Schrefler BA (2014) Three phase flow dynamics in tumor growth. Comput Mech 53(3):465–484

30. Sciumè G, Ferrari M, Schrefler BA (2014) Saturation–pressure relationships for two-and three-phase flow analogies for soft matter. Mech Res Commun 62:132–137

31. Sciumè G, Santagiuliana R, Ferrari M, Decuzzi P, Schrefler BA (2014) A tumor growth model with deformable ECM. Phys Biol 11(6):065004

32. Ulm F-J, Coussy O (1995) Modeling of thermo-chemo-mechanical couplings of concrete at early ages. J Eng Mech (ASCE) 121(7):785–794

Part III
Ageing in Infrastructure

The Relevance of Ageing for Civil Infrastructure: The Profession, the Politics, the Classroom

David A. Lange

Abstract The work of civil engineers and researchers has strong impact on public policy for infrastructure funding. This talk considers several professional, political and education-related thrusts. First is the challenge of communicating about intrinsically qualitative matters with metrics that support rational decision making for spending public funds. Second, it is necessary to confront political realities when trying to influence infrastructure-investment policy in free and democratic societies. Third, ageing of infrastructure is complex, and it demands new knowledge to address technical, economic and political aspects. Fourth, the new knowledge and expanded skills expected of future engineers are driving the reform of curriculum and pedagogy. This talk revels in the opportunity to marry technical facts with political commentary in the spirit of provoking thoughtful debate and discussion.

Keywords Public policy · Infrastructure funding · Sustainability Education reform

1 Introduction

This conference is entitled *The First International Conference on Ageing of Materials and Structures*. It is the first of what we hope will be a long and lasting conference series. Most of us in attendance are academic researchers and engineers who study a wide variety of technical issues related to material deterioration. Indeed, we are part of a long tradition of scientific research that aims to expand our knowledge of materials so we can improve the quality of infrastructure, achieve longer service life and possibly do it all at lower cost. So, it is right and fitting that the bulk of our conference is focused on technical topics and the latest experimental research results.

D. A. Lange (✉)
University of Illinois, Urbana, IL, USA
e-mail: lange@illinois.edu

© Springer International Publishing AG 2018
K. van Breugel et al. (eds.), *The Ageing of Materials and Structures*,
https://doi.org/10.1007/978-3-319-70194-3_5

This talk is a bit different. I was asked to think about the relevance and urgency of ageing issues regarding American infrastructure. What is the extent of the ageing-infrastructure problem, the money involved and the means to solve the problems? The *relevance* is huge and expansive as our society spends billions of dollars (or euros) on infrastructure. It is a goal of a free and democratic society that we work on behalf of the citizens who fund those expenditures so as to achieve the highest possible value for the benefit of all. The public-policy question is simple at its core: "How can we best spend our resources on infrastructure to achieve the highest benefit?" The implementation, of course, is complicated and controversial because everyone has a different answer to that question.

We as civil engineers and researchers have a unique and essential role. It is a role we do not often realize we play, as we may be quite content to stay within our closed academic community. But make no mistake—others are looking to us for answers. Our research sponsors are curious about our research for a reason! They are often trying to answer that very basic question—how can we best spend public money on infrastructure?

My strategy in this talk is to develop four lines of argument about the relevance of ageing infrastructure. First, I want to make a few comments about how we face very hard problems in describing the condition of infrastructure because it is a complex matter affected by material deterioration and structural performance that can be redefined as we learn, for example, more about seismic loads. It is never easy to transition from neatly *quantitative* parameters of scientific work to complex *qualitative* matters, such as infrastructure condition, the value of infrastructure for national competitiveness, costs of inadequate infrastructure and the cost for repair of complex systems. Second, I want to comment about how infrastructure-spending decisions are made in the United States, the level of infrastructure funding over the years and how infrastructure is a hot potato in US politics. How one communicates and persuades requires effective communication of infrastructure priorities targeted at the layman (i.e. politician) who may not be as interested in the technical story as he is in the political ramifications of infrastructure spending. Third, for the past ten years, we have seen the rise of *sustainability* as a powerful buzzword across all aspects of civil engineering, and, indeed, across the global economy. Sustainability is a critical theme that touches on repair, rehabilitation and reuse of infrastructure. Civil engineers naturally embrace sustainability because it is a central principle of responsible engineering practice. And fourth, the technical challenges of ageing infrastructure represent an opportunity for engineering educators. Education reform —happening in both North America and in Europe—can improve the effectiveness of higher education and elevate the technical skill set associated with infrastructure. The curriculum needs reform to address emerging topics, such as forensic studies, repair materials and service-life prediction.

2 The State of Infrastructure

2.1 Condition Survey and Infrastructure Inventory

Translating infrastructure condition into meaningful quantities is an ongoing challenge for agencies responsible for highways, airports, railroads and ship transportation. The practice of conducting condition surveys emerged as a matter of necessity for agencies overseeing diverse and geographically distributed facilities. Today, all US states have established practices to survey the condition of infrastructure, although uniformity among the states is not perfect. The federal government has made great strides in building national inventories of infrastructure. There is a myriad of inventories maintained at the national level: bridges, roads, railroad-grade crossings, airport runways, dams, pipelines etc.

The purpose of infrastructure inventories is slowly maturing as a tool for scheduled maintenance activities, and, as such, they gain value for budgeting activities. The procedures for inventories are increasingly defined by higher levels of government, and today the federal level is often specifying which parameters are to be measured and how they are to be measured in a consistent way.

Repair professionals have developed a methodology for condition surveys that can be applied to private and public infrastructure. Consultants are often engaged to give advice about repair or replacement of structures and pavements, and there are now ample tools for standardising the initial surveys required to assess the options for repair.

2.2 ASCE Infrastructure Report Card—"The Bridge is Falling!"

The American Society of Civil Engineers (ASCE) has been an advocate for the profession since its founding in 1852. With a mission to advance professional knowledge and improve the practice of civil engineering, ASCE strives to be a focal point for the transfer of research results and technical policy. In 1998, ASCE published its first Infrastructure Report Card (Fig. 1), taking a somewhat controversial approach to communicating the dire need for repair and maintenance. The idea of grading our nation's infrastructure did not originate with the ASCE, but came in 1988 from a presidential commission created to report on the state of US infrastructure. The presidential commission assigned an overall grade of C, and the title of their report "Fragile Foundations: A Report on America's Infrastructure" hinted at the shaky state of US infrastructure. ASCE has re-worked their analysis every few years, always arriving at a grade close to D. The latest report in 2013 observes a slight rise to D+. The report card comments on a wide swath of infrastructure categories, estimating, for example, that one-ninth of the nation's 600,000 bridges are structurally deficient.

Fig. 1 The ASCE infrastructure report card [8]

The impact of the ASCE report card has been striking. With every renewal of the report, the U.S. media gives fresh visibility to infrastructure quality. The report resonates well with the public as citizens have their own complaints about the poor quality roads. Tragic accidents have been attributed to decaying infrastructure. Perhaps one of the most publicized failures occurred in 2007 when the dramatic rush-hour collapse of the I-35 W bridge over the Mississippi River in Minneapolis caused the death of 13 people.

The ASCE report card is now familiar to all state and federal legislators, and over time it seems to have gained credibility as a touchstone for political posturing. President Obama made famous reference to the report card in his 2014 State of the Union address, using the speech as a launching point for new spending on infrastructure. Since the economic recession in 2008–2009, the U.S. government has asserted a variety of economic stimulus initiatives, increasing the flow of federal dollars in ways designed to promote economic stability. The 2014 emphasis on infrastructure is positioned as another spending plan that increases economic activity in a broad way, while seeking to achieve the broad and popular goal of improving infrastructure quality.

2.3   Contrary Viewpoints—"The Bridge is not Falling"

Not everyone thinks the ASCE report card is a good thing [1]. Many observers believe ASCE is playing politics with their report card. The methods for measuring infrastructure condition are extracted from national inventories, but the manner of interpreting the data is thought to be selective for the purpose of drawing attention to infrastructure spending. Infrastructure spending has been relatively steady for many years. Total public construction spending has varied between 1.7 and 2.3% of GDP for the last 20 years, according to the U.S. Census Bureau. By the Congressional Budget Office's slightly different measure, infrastructure spending has been between 2.3 and 3.1% of GDP since 1956 [2]. Another source estimated that the U.S spends about 3.3% of its federal budget on infrastructure, while in Europe the comparable figure is 3.1% [3]. The fiscal conservative flinches when ASCE trots out its final conclusion that $2.7 billion is needed to bring the infrastructure up to par when the current annual spending on infrastructure is in the $700 M range.

My goal today is not to parse out the debate between ASCE and its detractors, but to illustrate the clear point that civil engineers DO make a difference in the setting of public policy. Whether we want the role or not, we are a profession that holds the responsibility for infrastructure and seeks to deliver high value for maximum public benefit.

3   Making the Case for Strong Investment in Infrastructure

3.1   Infrastructure Spending in the United States

In 2013, the U.S. federal government spent $3.7 trillion. It is not easy to dissect the budget [I'll leave that job to the professionals!], but there are some pertinent statistics available to the common man. One measure of infrastructure investment is the budget of the Department of Transportation (although, admittedly, not all DOT money is spent on new construction or repair). The DOT budget has been reasonably stable since the 1980s at 2.0% of the federal budget. For 2013, the DOT budget was $750 M. The allocations within the DOT budget vary from year to year, and certainly there is a political element in how certain themes are featured as *programmes de jure*. For example, in 2010 the U.S. made a $1B initiative for high-speed rail in its annual spending plan above the previous year's spending level. In another case, a $4B Infrastructure Innovation Fund was established to stimulate projects with high priority.

Other sources suggest that actual expenditures on infrastructure have lagged in recent years [4]. Tracking expenditures in different ways suggests that actual

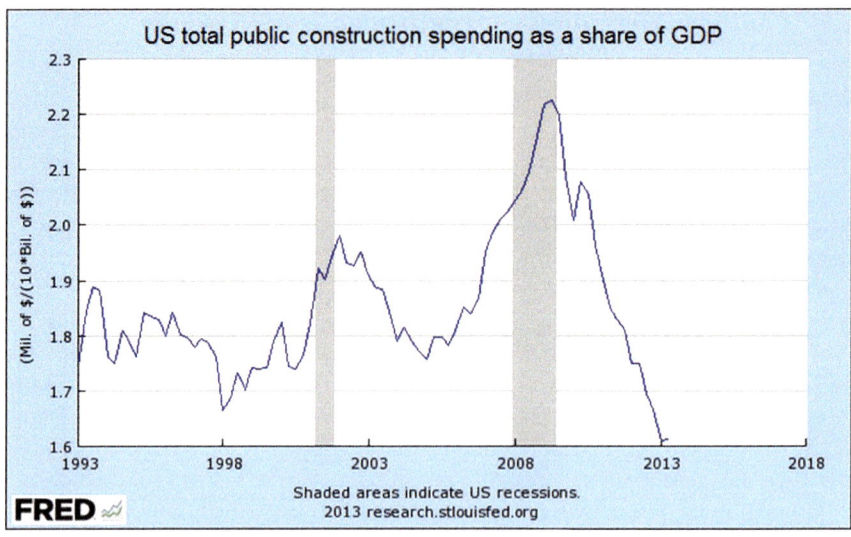

Fig. 2 US infrastructure spending [4]

spending on construction has fallen as shown in Fig. 2. The broad explanation for the contraction of spending, even as federal budgets are maintained, is that states and local governments are the biggest part of the story. Even federal dollars funnel through the states for the vast majority of spending on roads, highways and bridges, and the states have pulled back on spending since 2008 as a result of the economic downturn and requirements to balance state budgets. For example, California's transportation spending declined by 31% from 2007 to 2009 while Texas's fell by 8%.

3.2 National Competitiveness is Dependent on Infrastructure

High-quality infrastructure is a requirement for an efficient and effective modern society. The quality of infrastructure is one of the clearest differentiators between third-world and first-world economies. The World Economic Forum developed an index system for global competitiveness [5]. The U.S. comes out as seventh on their list of 144 nations in terms of competitiveness, and the nations ahead of the U.S. include European countries such as Switzerland, Finland, Sweden, Netherlands and Germany, along with Singapore. More interesting than the simple national ranks is the methodology used by the WEF. The competitiveness index includes 12 pillars, of which infrastructure is the second. [The 12 pillars are institutions, infrastructure, macroeconomic environment, health and primary education, higher education and training, goods-market efficiency, labor-market efficiency, financial-market development, technological readiness, market size, business sophistication and innovation.]

The infrastructure pillar represents fully 25% of the WEF global competitiveness index—a major statement about its importance. Half the infrastructure score comprises traditional infrastructure categories like roads, railroads, airports and ports. The other half comprises electrical and telephone infrastructure. Every country is scored, and problems are identified. For India (59th on the index ranking), the greatest problem is the inadequate supply of infrastructure. For the U.S. (again, seventh on the index ranking), the greatest problem is inefficient government bureaucracy, and infrastructure is near the bottom of the list.

4 Sustainability—a Guiding Principle

4.1 The Rise of Sustainability

Sustainability has been a popular emphasis across the civil-engineering profession for more than 10 years. Many conferences have been held with the word "sustainable" prominently featured in their titles. In the US, the American Concrete Institute reflected this trend when they established a new technical committee on sustainability in 2008 that quickly drew more than 100 members. In 2009, the Concrete Joint Sustainability Initiative was established, bringing more than 30 construction and material-trade associations together in a common cause. Sustainability, from the concrete-industry viewpoint, represents a broad set of issues including environment, economy and society. Our industry has emphasized recycling and the reuse of concrete materials as one of the most sustainable practices.

Ageing is relevant to sustainability because long service life is the best way to make concrete construction more sustainable. Ageing has many attributes, but materials durability is among the most threatening. Durability of concrete materials is an attractive focus because our current knowledge enables us to achieve very long service lives—perhaps beyond 100 years—if adequate materials, construction technology and reasonable maintenance practices are employed. Concrete is inherently a durable material unless economic compromise, poor workmanship or inadequate design leads to high permeability, cracking or overloading. Obsolescence is another factor that is hard to predict because needs and requirements can change in unpredictable ways, rendering worthless an otherwise serviceable structure.

4.2 The Future of Sustainability

Sustainable infrastructure demands commitment to principles of recycling, rehabilitation and reuse. New construction gets the headlines, but repair engineering is perhaps even more important for improving sustainability practices. The repair community has made tremendous strides over the past 20 years. The profession has elevated the visibility of repair, expanded the professional organizations and advanced the state of the art by creating codes and standards to ensure the use of best practices across the industry. For example, members of the International Concrete Repair Institute (ICRI) and ACI have worked together to create the ACI 562-13 *Code Requirements for Evaluation, Repair, and Rehabilitation of Concrete Buildings*. ACI, RILEM and other organizations have supported committees to address repair-related issues such as NDT, strengthening of existing structures and repair methodology. Further development of repair methods, standards and codes will come in the future to better equip the profession to respond to challenges of ageing infrastructure.

Sustainability as a rallying call seems a bit "old in the tooth," although these kinds of trends have long-lasting tails. I sense there is rising enthusiasm for the word "Resiliency" as the next big theme across the construction industry. Resiliency captures a sense of responsiveness, readiness and the ability to recover from disaster. Ageing is relevant also to the new theme of resiliency because infrastructure has to deliver performance at a level defined by the original design, and ageing is a threat that degrades performance over time. So, whether the focus is sustainability or resiliency, attention must be paid to the ageing of infrastructure.

5 Engineering Education Reform

5.1 How Did Our Education Model Become Stagnant?

I have been a faculty member at the University of Illinois for more than 20 years. For 6 years of that time, I served as the Associate Head for Undergraduate Affairs, and in that role I oversaw an undergraduate program with 800 students. During those years, I studied how our curriculum had evolved over the many years, and it was sobering to observe just how little change had occurred within my lifetime. I went to college in the 1970s, and our curriculum (course titles, credits and sequencing) at the time was actually quite similar to the University of Illinois in the 2010s. Almost course for course the same! My study drew me to look back as far as the archives would take me. The civil-engineering curriculum was very different 100 years ago. In the late 1800s, about a third of the curriculum was some aspect of surveying, whereas, in 1996, we ceased teaching the single remaining surveying course at UIUC. There was a large shakeup in engineering education in the years around WWII. After WWII, it became clear that technical pre-eminence was essential to national security, and there was a sense of urgency to build a strong educational system with a new commitment to the research enterprise. During 1940–1960, we saw adoption of the "Engineering Science Approach" that increased science and math content considerably and decreased technician-training aspects of the curriculum. But if you compare the today's curriculum with that of the 1960s, you'll see relatively little change. I believe that one of the strongest factors that led to stability and uniformity (and stagnancy) of U.S.-engineering curriculum is the accreditation processes used by universities and colleges. The most common accreditation body used by U.S. civil-engineering programs is ABET. While the expressed attitudes at ABET seem open to curriculum reform and innovation, the highly structured accreditation process remains an imposing barrier to change.

5.2 Education Reform in the US and Europe

Engineering educators are in the early stages of a wave of change that may emerge as the greatest since WWII. There is broad awareness that it is time to revisit the old education model [6]. Bold innovations are percolating through higher education, and much of it is driven by the desire to more effectively reach today's students who have grown up with different expectations shaped by their experience with computers. These Internet Age students have higher expectations with regard to the immediacy, interactivity and impact of course materials. Educators have been adopting innovations for course management, video, models and powerful computational tools like MATLAB. Online education is becoming mainstream, and the possibilities for archiving lecture material, pacing delivery to meet student expectations and self-study exercises has affected on-campus instruction as much as off-campus. Pedagogy is being taken more seriously, and educators are trying to understand learning processes so that teaching can be more efficient with better outcomes. Taken in sum, the wave of innovation is profound, and engineering educators are eagerly challenging assumptions about traditional curricular design. For example, *Design* is traditionally an activity for upper-level students who have taken all the science prerequisites. No longer. Design is being introduced at the freshman level as a motivator and as a way to convey context for the detailed course material yet to come. As another example, educators seek ways to better embrace the liberal arts. Feedback from the profession trumpets the need for broad communication skills and appreciation of the human condition in a global economy. Certainly our future demands different skills and knowledge than 50 years ago. Merely mentioning the advances of information technology and biological engineering is enough to persuade one that engineering-curriculum content needs to be responsive to changing needs.

ASCE has fostered debate about engineering education through its *Raising the Bar* committee reports. The premise is that engineering, as a profession, needs to

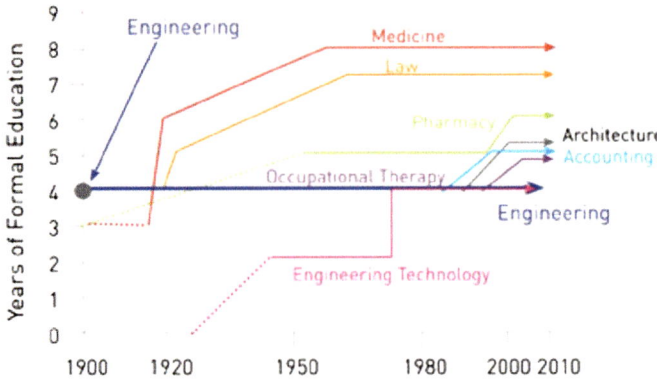

Fig. 3 Educational requirements for professions [9]

exert greater expectations for training as the needs of the profession rise. As shown in Fig. 3, other professions have "raised the bar" for training to enter their profession and earn licensure. States that hold responsibility for licensure are considered to hold the key to institutional change, and, over time, I think we will see the master's degree rather than the bachelor's degree used as the standard entry-level preparation.

Europe has wrestled with their own set of engineering-education challenges. The discussion is focused on uniformity across the EU and transferability of educational credentials. There is an initiative embodied in the Bologna Process that seeks to ensure comparability in the standards and quality of higher-education qualifications. To date, 47 nations are participating, indicating that a wide consensus has been established, and important steps are being taken to examine engineering education and its future. While the emphasis of the Bologna Process is international cooperation, the future of such dialog will reach curriculum, pedagogy and professional preparedness [7].

Topics like ageing beg for greater professional training and tools. Ageing is not a simple, one-dimensional, controlled lab experiment, fully explained by simple math, simple chemistry and simple physics. Rather, ageing is a complex area of study that requires synthesis, multidisciplinarity, design and problem-solving skills, economic sophistication, political savvy and skills of communication and persuasion. These requirement have profound implications for the education of future engineers.

This conference on Ageing of Materials and Structures, to me, is a call for professionals who may have seen themselves comfortable in "scholarly silos" to achieve greater awareness of, and enthusiasm for, their role in impacting public policy. We may enjoy the technical inquiry of the laboratory, but we are called to have broader impact outside the laboratory.

6 Summary

Ageing of infrastructure involves technical challenges, but the engineer and the researcher are also called to impact public policy that supports the high quality-infrastructure that benefits society. In addition, I cannot help but highlight the need for education reform to better meet the changing needs of the profession. My talk today seeks to make several points:

- Communicating ideas about ageing of infrastructure requires the ability to transform qualitative information (e.g. infrastructure condition) into quantitative metrics. The scientific knowledge underpinning such metrics needs to advance.
- Infrastructure quality contributes to national competitiveness, and, as such, investment in infrastructure is critical to economic vitality.
- Infrastructure spending has been relatively stable over time, but may shift topical focus under political pressure. Structural and long-lasting increases in infrastructure spending have been elusive, even as professional groups lobby in more sophisticated ways.
- Scientists and engineers have a role to play by informing the decision-making process that leads to public policy for infrastructure spending.
- Ageing is a complex topic that involves multidisciplinarity, design and problem-solving skills, economic sophistication, political savvy and the ability to synthesize and communicate all of these factors.
- Ageing connects well to other current themes in civil engineering: sustainability, resiliency and repair. All of these themes have implications for educating the next generation of engineers.
- Ageing and other rising needs of the engineering profession call for educators to update curriculum and reform pedagogy. Today's challenges cannot be met by educational approaches that have changed little over the past 50 years.

Acknowledgements I am grateful to the VTT-Fulbright Program for its support of my activities during my sabbatical leave in Finland. I appreciate the collaboration and discussions with my colleagues at VTT, Espoo, Finland who have contributed insights for this paper.

References

1. Soltas E (2013) Bloomberg view: the myth of the falling bridge. http://www.bloombergview.com/articles/2013-04-08/the-myth-of-the-falling-bridgettp://webmineral.com/data/Cryptomelane.shtml
2. Peterson SJ (2009) U.S. infrastructure spending: how much is enough?, Urban Land
3. Gregory PR (2013) Infrastructure gap? look at the facts. we spend more than Europe, Forbes. http://www.forbes.com/sites/paulroderickgregory/2013/04/01/infrastructure-gap-look-at-the-facts-we-spend-more-than-europe/

4. Pethokoukis J (2014) Actually, America doesn't have a trillion-dollar infrastructure crisis. American Enterprise Institute. http://www.aei-ideas.org/2013/11/actually-america-doesnt-have-a-trillion-dollar-infrastructure-crisis/. May 6
5. Schwab K (2012) The global competitiveness report 2012–2013, World Economic Forum
6. National Academy of Engineering (2005) Educating the engineer of 2020: adapting engineering education to the new century
7. Engineering Council (2014) Bologna process. http://www.engc.org.uk/education–skills/bologna-declaration
8. ASCE (2013) Infrastructure Report Card. http://www.infrastructurereportcard.org/grades/
9. American Society of Civil Engineers (2014) Raise the bar. http://www.raisethebarforengineering.org

Study on Triage for Deteriorated Concrete Structures by JSCE-342

Shinichi Miyazato, Takashi Yamamoto and Ryousuke Takahashi

Abstract In Japan, with the ageing of both structures and society, a social system should be constructed that reasonably maintains and manages safe concrete structures. Therefore, a Subcommittee on Priority in Maintenance and Management of Deteriorated Concrete Structures (JSCE-342) was organized to study schemes for determining the priorities during inspections and repairs. Four working groups investigated environmental action, performance evaluation and inspection to develop a priority-determination system. This paper outlines a midterm report; a maintenance procedure for structures based on priority determination is presented, and problems for the transition period from preemptive maintenance to preventive maintenance are identifies.

Keywords Deteriorated concrete structures · Triage · Structural performance evaluation · Priority maintenance system

1 Introduction

In Japan, the number of structures that are older than 50 years is increasing rapidly as shown in Table 1, [1], which means that the construction materials used in these structures have also aged. Consequently, many administrators are increasingly burdened with maintenance expenses that exceed their annual budget for maintenance.

At the same time, the birthrate in Japan has decreased rapidly. Figure 1 shows the time-dependent changes in Japan, in comparison to the UK and France. Hence, the

S. Miyazato (✉)
Kanazawa Institute of Technology, Nonoichi, Japan
e-mail: miyazato@neptune.kanazawa-it.ac.jp

T. Yamamoto
Kyoto University, Kyoto, Japan

R. Takahashi
Akita University, Kofu, Japan

© Springer International Publishing AG 2018
K. van Breugel et al. (eds.), *The Ageing of Materials and Structures*,
https://doi.org/10.1007/978-3-319-70194-3_6

Table 1 Ratio of infrastructure over 50-years old in Japan

	Road bridges (%)	River facilities (Gates etc.) (%)	Sewage ducts (%)	Piers (%)
AD2006	6	10	2	5
AD2016	20	23	5	14
AD2026	47	46	14	42

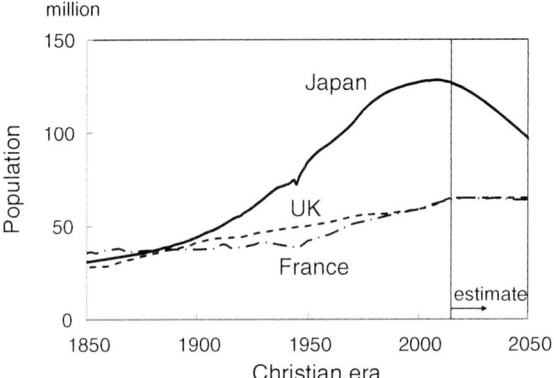

Fig. 1 Time-dependent changes of population in Japan, UK and France

Table 2 Members of JSCE342

WG1	○ J. Tomiyama, ● H. Minagawa, T. Ishida, H. Itoh, Y. Sakoi, T. Shiina, H. Takeda, H. Tanaka, J. Tamai, N. Nomura, S. Miyamoto
WG2	○ R. Takahashi, ● A. Kamiharako, ● Y. Mikata, Y. Koda, I. Kuroda, A. Sato, Y. Tanaka, T. Nina, T. Matsuo, T. Miki
WG3	○ K. Watanabe, H. Kano, H. Kawashima, T. Ohkubo, M. Sakurada, A. Shiba, S. Tanaka, K. Hara, T. Hara, Y. Yokota
WG4	○ S. Miyazato, ● T. Yamamoto, T. Asai, T. Hagiwara, K. Haruta, H. Ito, T. Iyoda, E. Kato, Y. Kato, S. Kitahara, T. Koyama, N. Nomura, T. Ogasawara, N. Sasaki, T. Yamato

○ chief, ● sub-chief

elderly will constitute the majority of the rural population in the future, which will lead to a scenario where social infrastructure will no longer be required in any region.

With the ageing of both structures and society, a social system should be constructed that reasonably maintains and manages concrete structures safely for arbitrary lengths of time (not semi-permanent) on a limited budget. In May 2010, the Japan Society of Civil Engineering organized the Subcommittee on Priority in Maintenance and Management of Deteriorated Concrete Structures (JSCE-342) to study schemes for determining the priorities during inspections and repairs. The Chairman was S. Miyazato, and the Secretary was T. Yamamoto. Table 2 lists the members of the working groups in JSCE-342.

This paper outlines a midterm report [2] published in 2012.

2 Organization of the Working Groups

Figure 2 and Table 3 present the time-dependent changes in the performance of a concrete structure and how they relate to the organization of the working groups in JSCE-342.

First, the actual condition of degradation due to environmental actions received during a previous in-service period is investigated to check the safety and comfort of the present structure, as described in (1). It is then judged to determine if it satisfies the performance requirements, as described in (4).

Next, the safety and comfort of the future structure are checked. Therefore, the present performance is investigated, as described in (2). The strength of environmental actions or loads and the deterioration speed are considered, as described in (3), and the future performance is predicted and checked, as described in (5).

Finally, a systematic investigation is performed to determine countermeasures, renewal or dissolution that can be applied to many structures, as described in (6).

From the above discussion, the four working groups were organized, as detailed in Table 3.

Especially in this paper, activities about WG2 and WG4 are described considering to the relevance of ageing for civil infrastructures.

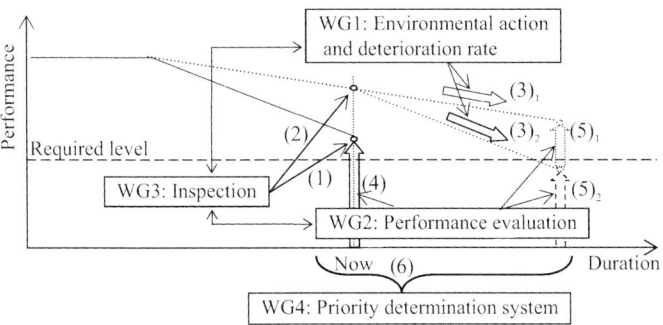

Note: Refer to Table 3 for (numbers) in figure.

Fig. 2 Time-dependent changes in performances and working groups

Table 3 Explanation of numbers in Fig. 2 and working groups

Number in Fig. 2	Explanation	Working group
(1), (2)	Performance of structure is inspected	3
(3)	Actions and deterioration rates are inspected and evaluated	1, 3
(4), (5)	Performance of structure is evaluated	2
(6)	Triage of maintenance and management is established	4

3 Activity for Structural-Performance Evaluation by WG2

3.1 Objective

The performance of an existing structure can be evaluated through a semi-quantitative method (grading) that uses the apparent deterioration and safety status for classification. Because this method is easy, the levels of many structures can be quickly evaluated. However, judging the suitability of countermeasures or renewal is difficult because the performance has not been quantified.

One method for determining the maintenance policy of existing structures is through performance-based design, where the ratio of the required performance to the remaining performance is calculated by numerical analysis. This can be clearly judged by using quantitative values. However, high levels of technology and investigation are necessary.

The advantages and disadvantages of these two methods were evaluated from the viewpoints of managers, constructors, inspectors and researchers. A scheme to introduce structural mechanics into the grading was also discussed.

Structural-analysis evaluation only using inspected data for grading was compared to grading evaluation. Examples for structures deteriorating from alkali-silica reaction (ASR) and frost attack are introduced in this section. Three-dimensional FEM analysis was used for structural-performance evaluation.

3.2 Example of Structures Deteriorating Because of ASR

Figure 3 shows a T-type pier of a road bridge. The head had cracks and reinforcing steel-bar breaks because of ASR. Rust leachate was not visible. Core samples were taken from the crack part, and the compressive strength was tested.

The maintenance outline for ASR of the road bridge [3] as shown in Table 4 was applied to the grading evaluation. According to this outline, the repair was judged as needed because the reinforcing steel bars were broken by ASR.

As a substitute for numerical analysis, Fig. 4 the FE mesh. The compressive strength of concrete was taken from the mean strength obtained for the grading evaluation. The reinforcing steel bars were introduced into the model as discrete members. The broken steel bar was introduced as discontinuous. Because rust leachate was not observed, the characteristics and adhesion of the steel bar were set as healthy. Figures 5 and 6 show the evaluation results. The failure mode was tensile bending. Localized cracks and a distorted distribution with right and left asymmetry occurred because of the break in the main reinforcement caused by ASR. The yield and ultimate loads were twice or more the design loads, which were the required values. Therefore, the structure was confirmed to have sufficient allowance.

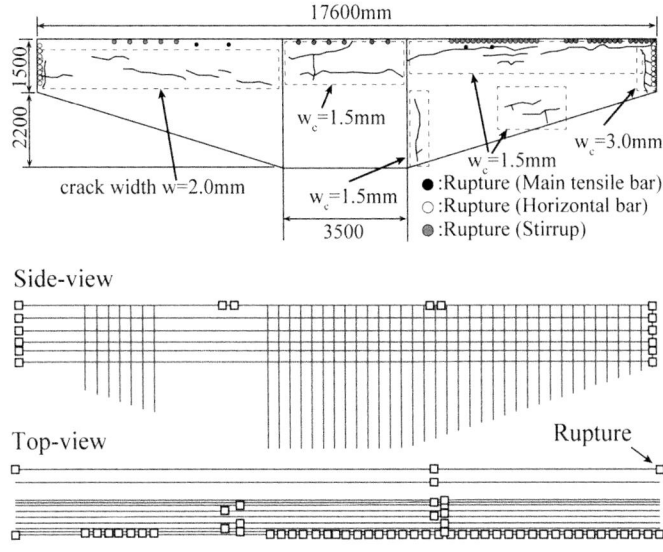

Fig. 3 Deterioration of T-type pier by ASR

Table 4 Damage needing repair

Case that is not yet repaired by crack injection method	Case that has been repaired by crack injection method
• There is a continuous crack whose width is more than 2 mm	• There is a continuous crack whose width is more than 2 mm, when the crack width before the previous repair adds to the crack width after its repair
• There is a continuous crack whose width is more than 1 mm at the member as shown in bottom figure	• There is a continuous crack whose width is more than 1 mm, if the crack width before the previous repair is unknown
• There is a bump whose height is more than 2 mm on the concrete surface at the crack point	• There is a continuous crack whose width is more than 1 mm at the member as shown in figure to the left
	• There is bump whose height is more than 2 mm on the concrete surface at the crack point

Crack by ASR

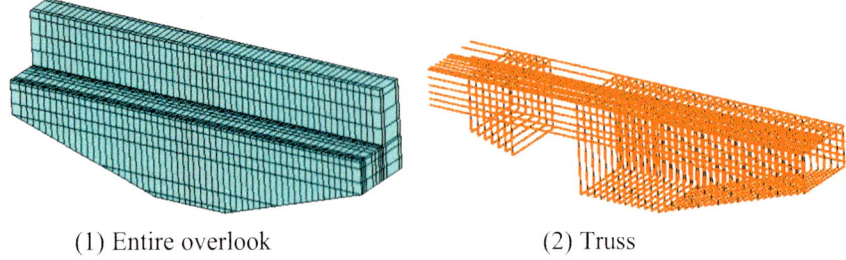

(1) Entire overlook   (2) Truss

Fig. 4 FE mesh

Fig. 5 Relation between load
and displacement

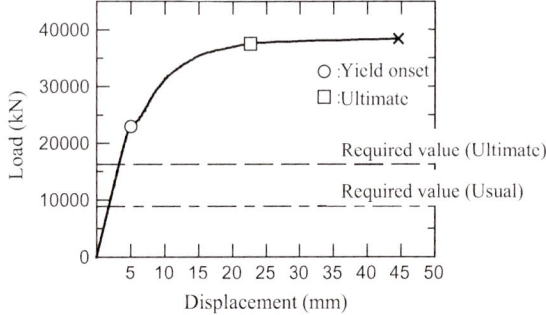

Fig. 6 Distortion contour

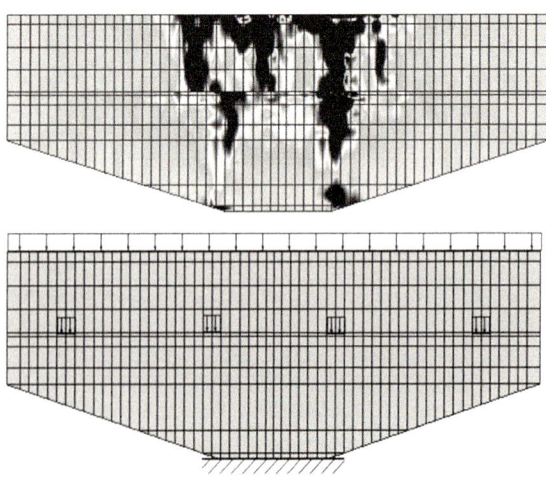

3.3   Examples of Structures Deteriorating from Frost Attack

Figure 7 shows an RC slab bridge. Cracks occurred in the lower part of the slab. The investigation only involved observation of the external structure.

The total assessment index of the road bridge [4, 5] was applied to grading evaluation. That is, a mark was given to every member depending on the damage situation, such as a slab crack or water leak. The deterioration level for each performance was then judged by totaling those marks.

As a substitute for numerical analysis, Fig. 8 shows the FE mesh. Because only visual information was available, the compressive strength, elastic coefficient and bond strength were unknown. Thus, the physical properties of the concrete were estimated from external observation and used in the numerical analysis. The deterioration levels of each member were classified into four phases based on the crack properties, i.e. the relationship between the outside grade and the relative dynamic modulus of elasticity was assumed, and physical properties of the concrete were estimated. Based on the environmental conditions and observation results,

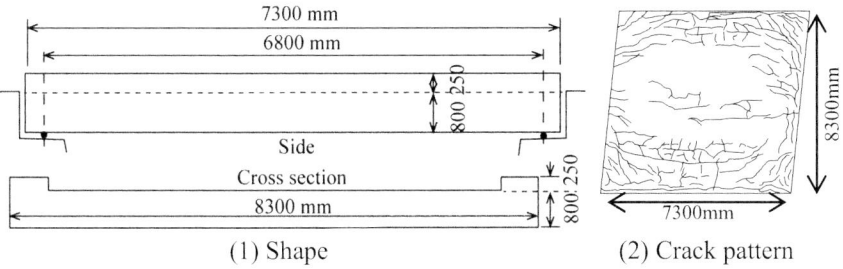

(1) Shape   (2) Crack pattern

Fig. 7 Deterioration of RC slab bridge from frost attack

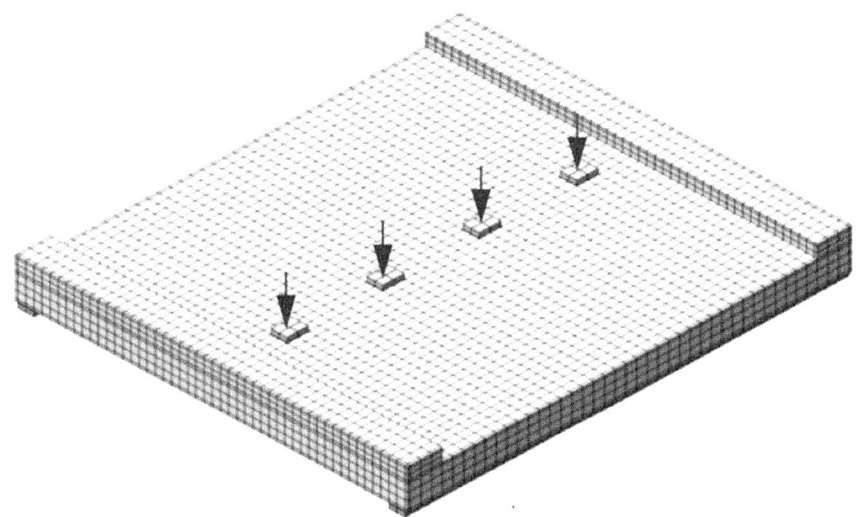

Fig. 8 FE mesh

Fig. 9 Relation between load
and center displacement

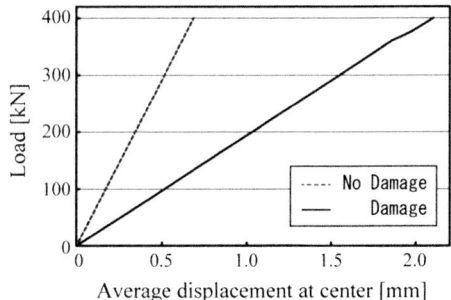

Table 5 Comparison of results for grading evaluation and numerical analysis

Performance	Method	No damage (A)	Damage (B)	Degradation ratio
Load-bearing ability	Grading	100 pt.	74 pt.	26% [(A) − (B)]
	Analysis	582 kN/mm	194 kN/mm	67% [1 − (B)/(A)]
Running stability	Grading	100 pt.	60 pt.	40% [(A) − (B)]
	Analysis	0.68 mm	2.1 mm	68% [1 − (A)/(B)]

the reinforcing steel bar was judged not to have corroded. Thus, for the discrete steel bars, only the bond strength decreased; the other strengths were rated as sound. The evaluation results are shown in Fig. 9.

Table 5 compares the results of grading evaluation and numerical analysis. The load-bearing ability and running stability decreased remarkably. The load-bearing ability decreased to 26% when the remaining performance was graded. On the other hand, the analysis results showed that it was 67%. The decrease in the grading evaluation became 40% when the running stability was evaluated in the same way. In contrast, the analysis result was 68%. Therefore, the grading evaluation was judged to be riskier than the results of the numerical analysis.

3.4 Summary

The accuracy of the numerical analysis was not high because the input data included many assumptions. However, the numerical analysis obtained results that considered the dynamic situation and could be compared with the required performance. For example, the main reinforcement in the T-type pier deteriorating from ASR did not greatly influence the structural performance, and the required performance was confirmed to be satisfactory. On the other hand, the RC slab deteriorating from frost attack more substantially affected the structural performance, compared to the grading evaluation.

In the present process of maintenance, it is thought that the scenario determining the priority is limited when structural-performance evaluation is utilised. The structures for which the performance evaluation is necessary are described as follows. That is, among the structures that require the detailed investigation to judge countermeasures, the preferred methods of countermeasure are not ascertained from only the detailed investigation results. For example, there are cases in which severe damage occurs, and it is necessary to verify the safety of the structure. Also, there are cases of structures in which the damage was confirmed without evidence of the type.

4 Activity for Priority Determination System by WG4

4.1 Objective

Preemptive repair of structures in dangerous conditions within several years of their discovery has been a major objective. However, for preventive maintenance, a scheme to prioritize maintenance of units over ten years is necessary in order to achieve the required performance, prevent accidents, avoid repair costs from exceeding the budget at any point in time (i.e. leveling life cycle cost (LCC)) and reduce the total LCC. Therefore, the present state and maintenance problems for an expressway, railway, port, harbor and local government were determined.

4.2 Results

When managing extremely important infrastructure such as the shinkansen, systematic maintenance is carried out at regular intervals according to the following procedure: inspection $\to$ judgment and evaluation $\to$ countermeasure. The accuracy of the predicted deterioration was improved by iterating the model over two to three cycles.

However, local government has gradually begun to carry out inspections of bridges. Improvements should be made to the inspection accuracy and utilization. If a maintenance-management system adapted to each administrator is established, an investment strategy for countermeasure, renewal or dissolution can be determined to draw up a budget.

After discussion, the necessity of a periodic maintenance cycle was confirmed. Table 6 presents problems for determining the priority.

Table 6 Problems for determining priority

Content	Problems
Deterioration of structure	• Evaluation unit of member and adequacy of representative value sampling • Appropriate inspection cycle interval for structure • Consideration of local environment (traffic volume, antifreeze agent) of structure
Difficulty of social factor	• Existence of substitute structure • Not all local governments prepare sufficient data

5 Conclusion

Because of the ageing structures and society, as described in Sect. 3.2, it will be difficult to simultaneously maintain all structures or members at the same level. Therefore, as shown in Fig. 10, selecting some structures or members for pre-emptive inspection and countermeasures based on engineering judgement is a reasonable approach. For example, the inspection frequency for a structure in a region with strong environmental impact can be increased. The structural perfor-mance should be analysed using input data that can easily obtained to judge whether the required performance level is satisfied.

Table 7 presents problems during the transition period from "preemptive maintenance" to "preventive maintenance." The deterioration of existing concrete structures should be investigated regularly; promoting rational maintenance is necessary.

Developing countries are rapidly developing their social infrastructure. Fur-thermore, several decades later, these countries will be faced with low birthrates and an ageing population. Therefore, in view of the actual conditions currently preva-lent in Japan, it is expected that developing countries will give priority to the establishment of management systems for the maintenance of concrete structures.

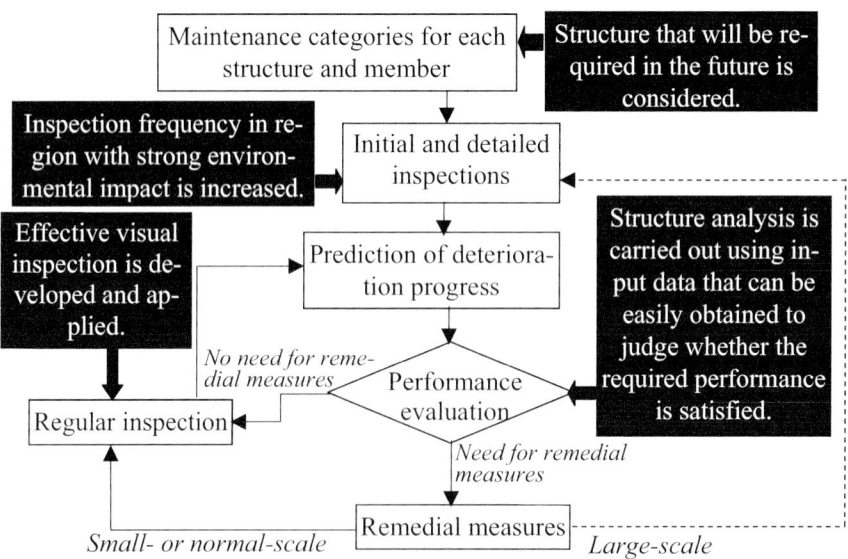

Fig. 10 Maintenance procedure for structures based on priority determination

Table 7 Problems for the transition period from preemptive maintenance to preventive maintenance	A grading technique with a semi-quantitative index should be developed to evaluate the structural performance of a large number of structures in the screening phase. Thereafter, a determination system should be established to identify structures whose structural performance needs to be quantitatively evaluated
	A more efficient maintenance system involving "inspection," "evaluation and judgment" and "remedial measures" will be required. After two or three cycles of the maintenance system, it should perform a self-assessment. Thereafter, a triage system suited to the requirements of each administrator should be established
	A reliable technique that can establish environmental impacts and deterioration rates easily, widely and quantitatively is necessary
	In accordance with "preemptive maintenance," visual inspection may be carried out for structures with obvious damage. In contrast, non-destructive inspection will be useful for structures without obvious damage in accordance with "preventive maintenance." A new inspection scheme needs to be established that considers the total costs

Acknowledgements This research was partially supported by the JSPS KAKENHI Grant Number 26289137, the Cross-ministerial Strategic Innovation Promotion Program from JST and the Strategic Research Foundation S1312006 from MEXT.

References

1. Ministry of Land, Infrastructure and Transport (2008) White paper on ministry of land, infrastructure and transport
2. Japan Society of Civil Engineers (2012) Midterm report of subcommittee on priority in maintenance and management of deteriorated concrete structures
3. Ministry of Land, Infrastructure and Transport (2003) Maintenance control summary for alkali silica reaction of road bridge (draft), Press release
4. National Institute for Land and Infrastructure Management (2007) Basic data collection manual of road bridge conditions (draft). Study on basic survey of a highway bridge conditions, Technical note, vol 381
5. National Institute for Land and Infrastructure Management (2008) Annual report of basic data on road structures in FY 2007, Technical note, vol 488

Ageing in Shallow Underground RC Culverts and Tunnels

M. Kunieda, X. Zhu, Y. Nakajima, S. Tanabe and K. Maekawa

Abstract The safety and serviceability of the concrete structures in service are probably affected by long-term ageing. Creep deflection of concrete, which depends particularly on the loading history and water content, as a phenomenon of ageing, has been proved to be an unavoidable problem. This paper aims to investigate the mechanism of the long-term excessive deformation of shallow RC box culverts, which have members of comparatively smaller thickness. The authors mainly focus on two points: the internal moisture state of concrete members and the increase in the vertical earth pressure acting on culverts caused by both uneven settlement of the foundation and RC structural deformation as a coupled action. To examine these effects, sensitivity analyses using multi-scale analysis and site investigation for deformational modes and steel strain are conducted. It is concluded that long-term ageing of shallow underground RC culverts and tunnels may be attributed to the synergistic effects accompanying risky delayed shear deformation owing to redistributed vertical soil pressure accelerated by the structural deformation associated with shrinkage and creep of concrete.

Keywords Aging · Long-term excessive deformation · Underground RC culverts · Internal moisture state · Differential settlement · Delayed shear deformation

M. Kunieda
School of Science, Aalto University, Helsinki, Finland

X. Zhu
Shimizu Corporation, Tokyo, Japan
e-mail: zhu@concrete.t.u-tokyo.ac.jp

Y. Nakajima · S. Tanabe
Tokyo Electric Power Company, Tokyo, Japan

K. Maekawa (✉)
Department of Civil Engineering, The University of Tokyo, Tokyo, Japan
e-mail: maekawa@concrete.t.u-tokyo.ac.jp

© Springer International Publishing AG 2018
K. van Breugel et al. (eds.), *The Ageing of Materials and Structures*,
https://doi.org/10.1007/978-3-319-70194-3_7

1 Introduction

1.1 Research Background

Ageing of concrete structures, which is an inherent feature of nature, is a time-dependent process. For concrete structures, creep deflection as a phenomenon of ageing, which depends particularly on loading history and water content, should be given full concern regarding the safety of structures.

Since the 1990s, the long-term monitoring of Tsukiyono Bridge's deflection has been periodically reported by Hata et al. [4], and the imperfection of design methods based upon the conventional linear creep law and shrinkage has been discussed. Recently, excessive deflections of cantilever PC viaducts have been reported worldwide by Bazant et al. [2] as well. Maekawa et al. [6, 11] point out two main causes of excessive deflection; one is the non-uniform thermodynamic state of moisture inside micropores and associated creep, and the other is the delayed average shrinkage of upper and lower flanges in time. The mechanism of the long-term deflections of PC viaducts has been made clear by considering these two factors.

On the other hand, underground facilities that are essential parts of the urban infrastructure are used for a wide range of applications—from small pipeline networks to large underground structures, including subway and highway tunnels. One of the maintenance problems of these underground structures in service is the long-term excessive deflections of shallow RC box culverts, which have members of comparatively smaller thickness. After a few decades of service, the top-slab deflection may exceed approximately three to ten times the prediction based on conventional design formulae. Large numbers of cracks at the inner surfaces of top slabs have been observed. This ageing phenomenon represented by excessive deformation and increasing cracks of unknown causes is a great concern in terms of future risk and serviceability.

1.2 Scope

Long-term excessive deformation and large cracks could be observed as a main phenomenon of concrete structures ageing. This paper aims to investigate the ageing mechanisms of the long-term excessive deformation of underground box culverts coupled with the soil foundation. Within the scope of this study, two points that have been neglected for verification of the long-term serviceability-limit state in design are focused on. One is the nonuniform internal moisture state of concrete members, which has been pointed out as one of the main causes of the long-term deflections of PC viaducts. The drying-creep deflection derived from internal moisture loss of concrete members is assumed to be negligible by the Japanese Road Association Code because moisture-containing soil consistently keeps

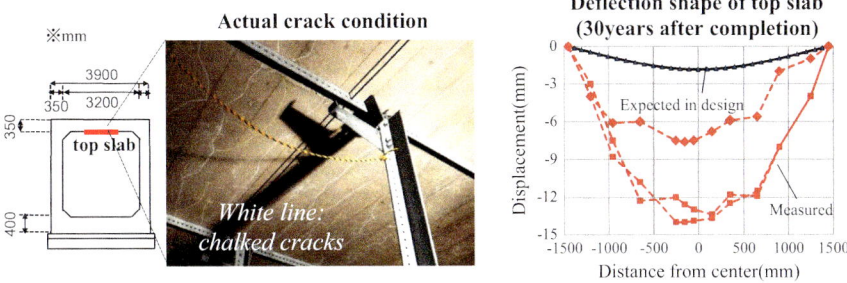

Fig. 1 Observed ageing phenomenon: long-term excessive deformation and crack condition

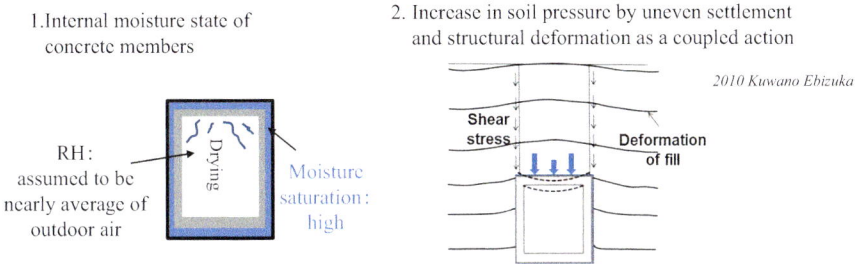

Fig. 2 Two focus points for investigating long-term ageing [5]

concrete members wet. In consideration of the actual ambient environment of the targeted RC culverts, the relative humidity (RH) of the space inside the culvert is thought to be close to the average RH of outdoor air, but the outer surface exposed to the soil foundation is kept wet. This implies a nonuniform moisture state in concrete members.

The other point is the increase in the vertical earth pressure acting on culverts caused by both uneven settlement of the foundation and RC structural deformation as a coupled action. In the case of a shallow underground structure whose over-lay is not greater than 10 m, a number of culverts carrying loads greater than the overburden soil have been reported [1, 3]. The increased vertical soil pressure on the top slab may accelerate long-term deformation, as well (Figs. 1 and 2).

2 Pre-analysis

2.1 Outline of a Multi-scale Integrated Model

To grasp the effects of the two focused points on long-term ageing phenomenon, the analytical system *DuCOM-COM3* [7] is used. This is a multi-scale analysis code

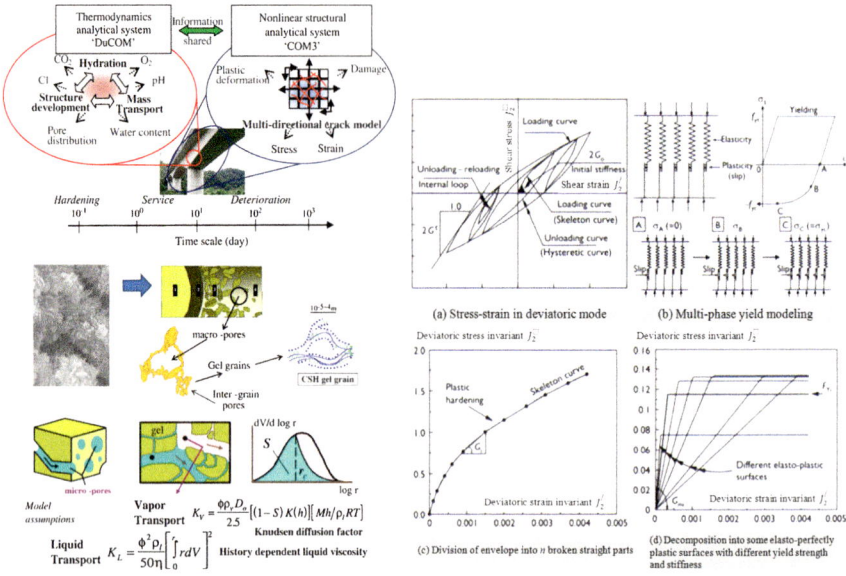

Fig. 3 Outline of the multi-scale integrated analytical system and constitutive model for soil

that links the thermo-chemo-physics platforms *DuCOM* [8], and *COM3* [9], as shown in Fig. 3.

DuCOM is an integrated thermo-hygral analysis model, which includes cement hydration in concrete mixture, micropore structure formation and mass transport models for concrete ranging from the 10^{-3}-to-10^{-9}-m scales of micropores, while *COM3* is a 3D finite element analysis platform for structural concrete with and without cracks. As a result, the linked system is capable of predicting changes in concrete material properties from casting to dismantling of entire structures and taking this material development into account for predicting the response of structural concrete. Through such integration, the long-term structural response of ageing under actual ambient conditions can be simulated in a more realistic manner. Figure 3 illustrates the code linkage for computing the nonlinear, time-dependent responses of reinforced concrete.

A nonlinear path-dependent constitutive model for soil is also used to meet the challenge of discussing soil-structure interaction. In this study, soil is mechanically idealized as an assembly of a finite number of simple, perfectly elastoplastic elements connected in a parallel pattern as shown in Fig. 3 [13, 12]. As each component is given different yield strengths of plasticity, which may reflect the grading of the sand particle size, all components subsequently begin to yield at diverse total-shear strains, which results in a gradual increase of entire nonlinearity. This analytical system can treat the time-dependent interaction behaviour of RC structures and soil in the same framework.

2.2 Pre-analyses for Evaluating Increased Soil Pressure

Pre-analyses targeting the trap-door test [5] were conducted for experimental verification of the vertical soil pressure owing to differential settlement. The overview of the test [5] is shown in Fig. 4. The FEM mesh and the soil properties are defined in reference to the experiment [5]. Joint interface elements are put on the boundary between soil elements and others, which allows free shear slip to reproduce the side layers of foundation.

Figure 5 shows the comparisons of experimental and analysis results. In general, *DuCOM-COM3* is capable of reproducing the increased soil pressure on the stable door due to the downward displacement of the movable doors. Not only the rapid increase in soil pressure with relatively small downward displacement (maximum soil pressure is observed at 0.5 mm of displacement), but also the shear plane, which has a strong relationship with the mechanism of increasing soil pressure, can be well simulated. In other words, this analytical system can be applied to simulate the phenomenon of the increase in vertical earth pressure acting on the culvert caused by uneven settlement, when the amount of uneven settlement is approximately known.

Fig. 4 Experimental condition [5] and half analytical mesh

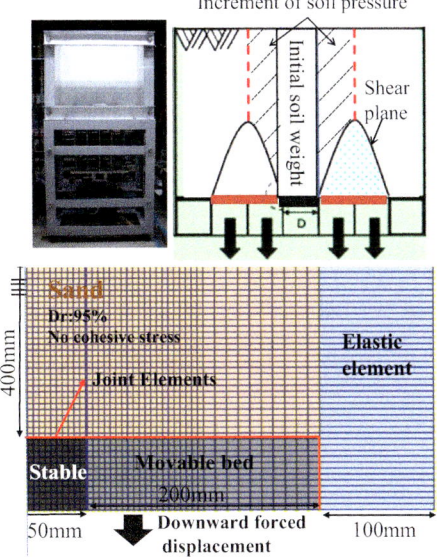

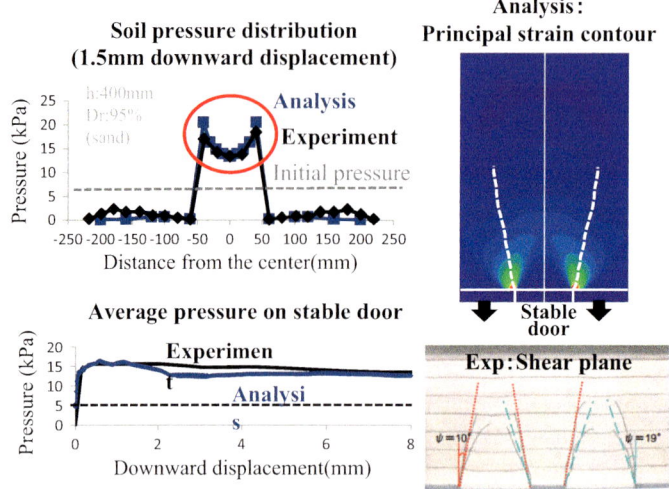

Fig. 5 Analytical and experimental results for change in earth pressure with downward displacement and shear plane

3   Verification of Main Causes of Ageing

3.1   Details of Four Ageing RC Culverts

In order to examine the effects of the two points (Sect. 1.2), a series of sensitivity analyses were conducted. Furthermore, analytical results and site investigation of deformational modes and steel strains obtained by destructive testing were compared for discussion of the mechanism for generating long-term excessive deformation. As for analytical targets that have exhibited long-term aging phenomenon, four RC culverts aged around 30–50 years were modelled and analysed. The details are shown in Fig. 6. The thickness of the concrete members are comparatively thin (25 or 35 cm), and the height of the fill above the culvert is relatively small, ranging 3.8–6.4 m. The backfilled materials, which may affect the increase in earth pressure, differ for the four culverts. Culvert F, for which specific data about soil properties is available (typical sand used for backfilled material), is the main focus of the sensitivity analyses.

3.2   Analysis Condition for Sensitivity Analyses

In reference to the structural details and material data, a half-mesh domain was created. For all analyses, the culverts are exposed to various environmental

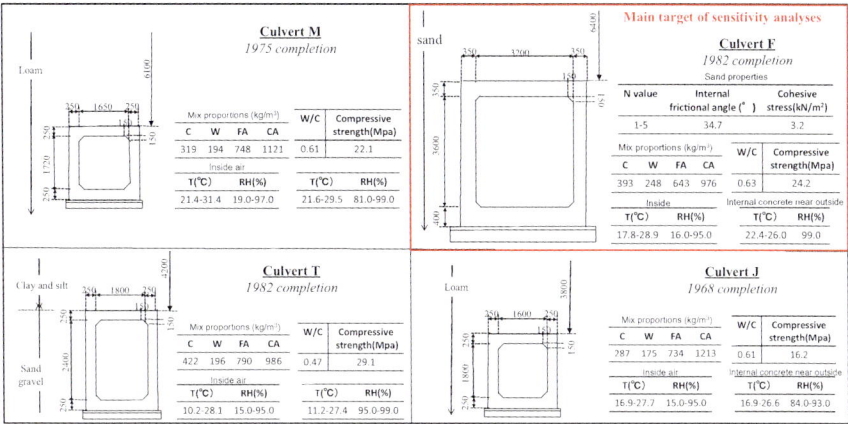

Fig. 6 Details of the targeted RC culverts (cross-section view and material properties)

conditions after 28 days of moisture curing. Construction-process complexity was eliminated for the sake of clarifying the influencing factors.

To figure out the effects of "increase in soil pressure due to uneven settlement" and "drying shrinkage, creep derived from the internal moisture state of concrete members" on the long-term deflection of the top slab, the following four cases are discussed; Case (I) no internal driving forces provoked by concrete drying, no uneven settlement; Case (II) consideration of uneven settlement, but no internal driving force; Case (III) consideration of internal driving force, but no uneven settlement; and Case (IV) consideration of all interactions. Thus, the difference between Case I and Cases II, III, IV respectively represent each effect described above (Figs. 7 and 8).

The relative humidity (RH) of the inside and outside boundaries of the culvert was kept at 99.99% from casting, which means that the internal moisture of concrete members was consistently maintained. The ambient temperature and RH of Case III and Case IV were maintained at the annual averages (inside culvert: 25 °C RH 60%; outside: 20 °C RH 99%).

Uneven settlement was reproduced by applying downward-forced displacement at the bottom of the surrounding soil. In Case II and Case IV, uneven settlement causing maximum increase in soil pressure was intentionally produced. That uneven settlement was applied over the two days immediately after completion in order to assume the severest condition, meaning the long drying-creep effect of the top slab. The real effect of the increase in earth pressure due to uneven settlement was somewhere in between the two extreme cases (Case I and Case IV).

Fig. 7 Long-term deflection at mid-span and deformational mode at current age

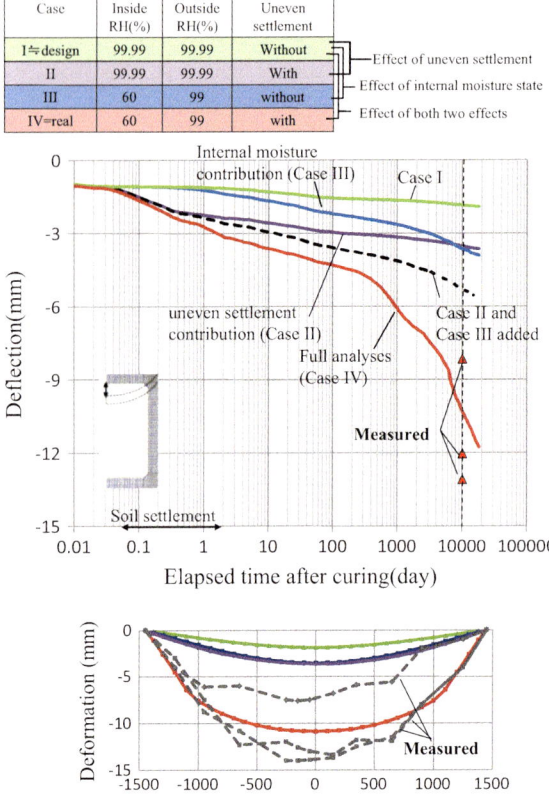

Fig. 8 Tension-steel strain at mid-span with average vertical soil pressure on the culvert

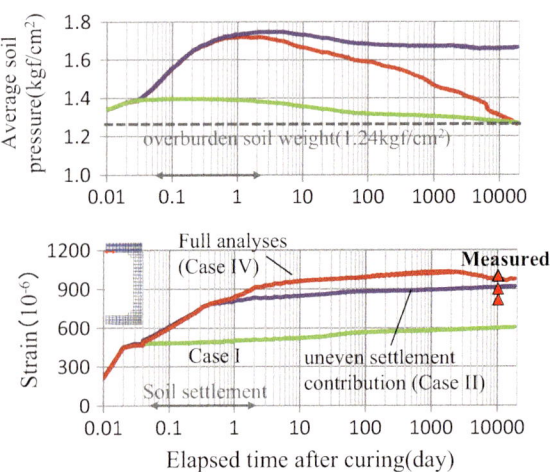

3.3   Analysis Results

Case II considering uneven settlement shows a small increase in deflection compared with Case I without uneven settlement. This is because the simulated steel strain grows with the increase in average vertical-soil pressure due to the uneven settlement during the early age. The simulation considering uneven settlement is capable of reproducing the measured tension-steel strains. This means the soil pressure acting on the culvert is also well simulated because the tension-steel strains reflect the external force mainly governed by the soil pressure. The simulated increment of average soil pressure is up to about 1.37 times the pressure of overlay soil mass. This value seems reasonable from both the calculated value (below 1.6-times soil-pressure increment) based on the suggested equation [5] and the other experiment using backfilled sand, which shows a 1.32-times increase in soil pressure [1].

As shown in Fig. 9, the internal moisture is gradually lost from the inner surface, and the drier the concrete members become, the faster the growth of compressive strain due to the drying creep. This effect on the long-term deflection seems to be limited because the greater part of the compression side is wet due to the moisture containing soil.

The analysis of simply adding these individual effects leads to underestimated deflections, but the full-coupling analysis shows good agreement with the measured results. The aging mechanisms can be considered to be the simultaneous and synergistic effect of two factors: first, the non-uniform internal-moisture state of concrete members due to water lost with ageing; second, the increase in the vertical earth pressure acting on culverts caused by both uneven settlement of surrounding backfill soil and RC structural-ageing deformation as a coupled action.

The time-dependent long-term deflection from 15 to 30 years of the full-coupling analysis also matches well the reported crack growth. The cause of this rapid increase in the long-term deflection after a few years is assumed to be the

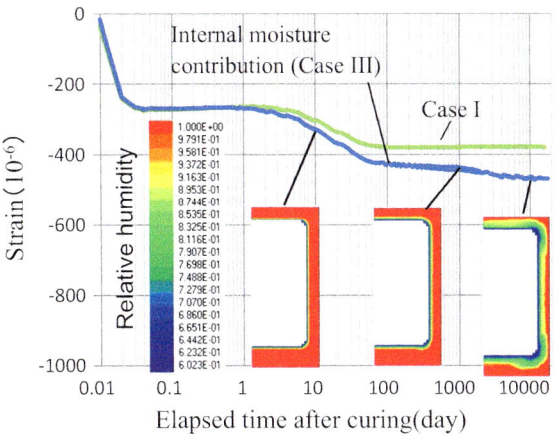

Fig. 9 Compressive steel strain at mid-span with internal RH contour figure

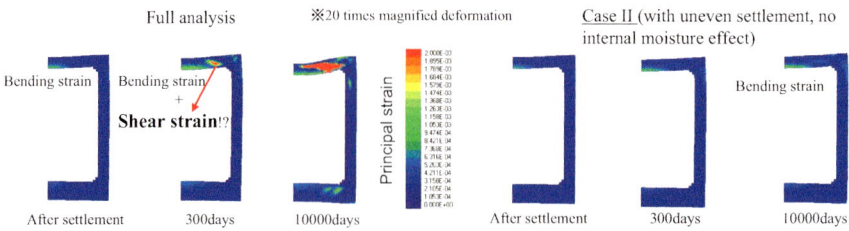

Fig. 10 Principal strain contour of full analysis (**left**) and Case II (**right**) based on half-mesh model

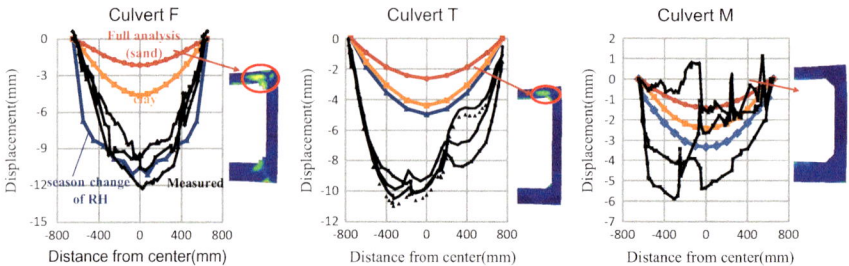

Fig. 11 Simulated deformational modes of current age with principal strain contour

delayed shear failure around the haunch indicated by the principal strain contour (Fig. 10). The delayed shear might be considered as a phenomenon of the ageing of concrete structures.

4 Ageing Mechanism of Accompanying Delayed Shear and Verification

4.1 Simulated Ageing Deformational Modes

To further investigate the ageing phenomenon of delayed shear associated with excessive deflection, the other three culverts were also modelled and analysed. The simulated deformation of the other three culverts was more or less underestimated, which is shown in Fig. 11. Since soil properties are not known in these cases, the authors assume backfilled dry sand. But, according to the construction records, there is the possibility that cohesive soils were used. If so, the higher soil pressure induced by the foundation subsidence might be simulated. Another uncertainty is the seasonal change of the culverts' indoor relative humidity, which has never been recorded. Thus, the annual average was used for simulation. However, the principal strain contours of culverts J and T, considering the uneven settlement and the internal moisture state, show shear–strain concentration near the haunch.

In addition, the cracking actually seen close to the haunch is located in the flexural compression region. This fact also indicates the possibility of accompanying delayed shear.

4.2 Investigation into Ageing's Cause and the Mechanism of Delayed Shear

Delayed-shear deformation might be a new point of discussion in terms of aged concrete structures. As a cause of accompanying delayed shear, the increase in the shear force acting close to the haunch should be first examined. Figure 12 shows the distribution of the vertical soil pressure acting on the culvert with the passage of time. The soil pressure at the middle is gradually released to about 60% of the overlay soil weight and the soil pressure near the haunch increases to approximately twice the initial soil weight. This redistribution of soil pressure is thought to be triggered by the deformation of the top slab because large deflection at the middle and small deformation around the haunch take place in analysis. This redistribution of soil pressure is also indicated by measuring the actual soil pressure on the culvert [1]. As a result of this redistribution, the shear force acting on the cross-section showing the shear deformation is slightly increased after 100 days to 1,000 days (Fig. 13). This increase in the shear force triggered by the structural-aging deformation could be one of the causes of accompanying delayed shear.

The other cause of delayed shear is considered to be the reduction of the shear capacity by both the higher sustained load close to the static capacity and the long-term drying shrinkage. It is reported that the shear capacity of the simple beam that bears the influence of drying shrinkage is reduced by up to about 85% of the static capacity under the sealed condition [10]. It is reasonable to consider that the shear deformation occurs because of the sustained load close to the shear capacity for a few years under the constant drying condition. Experiments to determine shear capacity under sustained loads near the capacity are required.

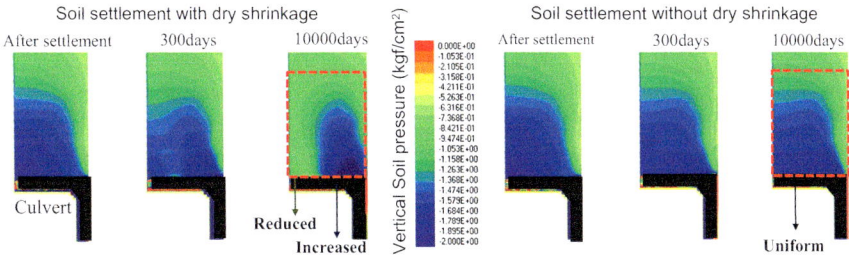

Fig. 12 Vertical direction contour showing distribution of soil pressure (culvert F)

4.3   Site Investigation into the Existence of a Delayed-Shear Crack

To verify the possibility that the actual RC culvert has shear deformation, a site investigation to carefully check for the existence of inner surface cracks near the haunch was conducted. Considering the bending moment acting on the culvert (Fig. 14), the cracks should be concentrated on the tension side, which means the centre of the top slab. The observed cracks were not only on the inner surface of the centre, but also near the haunch, which is intrinsically the compression side in flexure. The largest crack width near the haunch is 0.9 mm, which is more than three times the allowable value specified, even though the crack width at the centre is 0.3 mm.

It may be reasonable to hypothesise that the large cracks near the haunch are caused not by bending moment, but by delayed-shear cracks. Here, it must be noted that, unlike a statically determinate structure in air, the investigated soil-structure system is computationally under stability even after diagonal-crack propagation in the top slab. This is because the reduced member stiffness caused by shear cracks may lead to the re-distribution of soil pressure, and a newly produced stress flow is formed to bear the load-carrying mechanism as a result of the static soil-structure interaction.

Fig. 13 Average shear stress of one section accompanying delayed shear

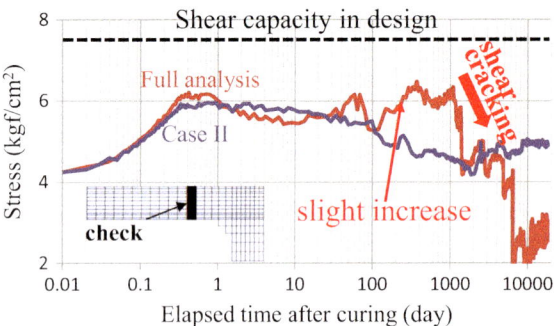

Fig. 14 Assumed crack area and crack site-investigation

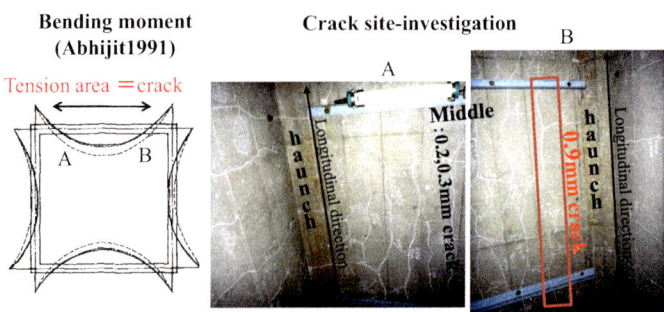

5 Conclusions

The following conclusions and views were drawn from both the thermo-hygral and mechanics analyses and site inspections:

(1) Soil settlement brings about the imposition of excessive overlay loads upon culverts that exceed the design specification equivalent to the dead weight of soil above the top slabs.

(2) The moisture gradient caused by the drying ambient states inside culverts accelerates long-term ageing of structures over several decades.

(3) Analytical results considering the interactive behaviour of structures and soil show that ageing phenomenon represented by excessive deformation accompanying delayed-shear deformation results in the redistribution of the vertical soil pressures triggered by structural deformation. This scenario is partially supported by site inspection to observe the large cracks around the haunch. Further investigation is required to upgrade the serviceability limit-state design of ageing structures.

Acknowledgements This study was financially supported by JSPS KAKENHI Grant No. 23226011.

References

1. Abhijit D, Bratish S (1991) Large-scale model test on square box culvert backfilled with sand. ASCE J. Geotech. Eng. 117(1)
2. Bazant ZP, Yu Q, Li GH, Klein GJ, Kristek V (2010) Excessive deflections of record span prestressed box girder. Concr. Int, ACI, 32(6)
3. Bennett R, Wood S, Drumm E, Rainwater N (2005) Vertical loads on concrete box culverts under high embankments. J Bridge Eng 10(6):643–649
4. Hata Y, Oonishi N, Watanabe Y (1993) Creep behavior of prestressed concrete bridge over ten years. In: Proceedings of the FIP symposium, pp 305–310
5. Kuwano R, Ebizuka H (2010) Trapdoor tests for the evaluation of earth pressure acting on a buried structure in an embankment. In: Proceedings of the 9th International symposium on new technologies for urban safety of mega cities in Asia, USMCA, Kobe, October 2010, CD-ROM
6. Maekawa K, Chijiwa N, Ishida T (2011) Long-term deformational simulation of PC bridges based on the thermo-hygro model of micro-pores in cementitious composites. Cem Concr Res 41(12):1310–1319
7. Maekawa K., Ishida T, Kishi T (2009) Multi-scale modeling of structural concrete. Taylor & Francis
8. Maekawa K, Chaube R P, Kishi T (1999) Modeling of concrete performance—hydration, microstructure formation and transport, London, E & FN Spon
9. Maekawa K, Pimanmas A, Okamura H (2003) Nonlinear mechanics of reinforced concrete. SPON Press

10. Mitani T, Hyodo H, Ota K, Sato R (2011) Discover and the evaluation of shear strength decrease of reinforced normal-strength concrete beams. Proc JCI 33(2):721–726
11. Ohno O, Chijiwa N, Suryanto B, Maekawa K (2012) An investigation into the long-term excessive deflection of PC viaducts by using 3D multi-scale integrated analysis. J Adv Concr Technol 10:47–58
12. Soltani M, Maekawa K (2015) Numerical simulation of progressive shear localization and scale effect in cohesionless soil media. Int J Non-Linear Mech 69:1–13
13. Towhata I (2008) Geotechnical earthquake engineering. Springer

Part IV
Ageing and Radiation

Radiation Shielding Properties and Freeze-Thaw Durability of High-Density Concrete for Storage of Radioactive Contaminated Soil in Fukushima

Sanjay Pareek, Yusuke Suzuki, Ken-ichi Kimura, Yusuke Fujikura and Yoshikazu Araki

Abstract In this research work, high-density concrete ($\mu = 4.71$ g/cm^3) using steel balls as aggregates and a normal concrete using recycled aggregates from the debris of the demolished concrete buildings of the earthquake-affected region have been evaluated for the radiation-shielding property for the radioactive contaminated soil in Fukushima. Two cylindrical model containers for storage of radioactive contaminated soil samples have been made using these two types of concretes. From the results of the experiments, it was demonstrated that the high-density cylindrical concrete container with a concrete-shield thickness of 100 mm can reduce radiation-dose equivalents emitted from radioactive cesium in contaminated soil by up to 90%. Good agreement was observed between the experimental and calculated-dose rate using Monte Carlo simulation (MCNP4C2 code) for two types of concrete for various shield thicknesses and measurement distances. Also, the freezing and thawing durability of high-density concrete was found to be superior to normal concrete using recycled aggregates.

Keywords Radioactive cesium · High-density concrete · Recycled aggregate
Shielding design · Freezing and thawing

S. Pareek (✉)
College of Engineering, Nihon University, Koriyama, Fukushima-ken, Japan
e-mail: pareek@arch.ce.nihon-u.ac.jp

Y. Suzuki
International Research Institute of Disaster Science, Tohoku University, Sendai, Japan

K. Kimura · Y. Fujikura
Fujita Corporation, Tokyo, Japan

Y. Araki
Graduate School of Engineering, Kyoto University, Kyoto, Japan

© Springer International Publishing AG 2018
K. van Breugel et al. (eds.), *The Ageing of Materials and Structures*,
https://doi.org/10.1007/978-3-319-70194-3_8

97

1   Introduction

After the disastrous accident at the Fukushima Daiichi Nuclear Power Plant in March 2011, a huge amount of radioactive material was dispersed into the air and contaminated a large land-surface area of Fukushima prefecture with radioactive cesium (^{134}Cs and ^{137}Cs). The government agencies have conducted vast decontamination activities throughout the region by scraping the ground down a few centimetres and collecting the upper layer of contaminated soil, along with organic waste from the surrounding vegetation [1]. The large volume of collected contaminated soil is temporarily stored in plastic bags (1-Tonne packs) and need to be stored safely for an unforeseeable time period. In order to solve this problem of storage of radioactive contaminated soil by a safe method, high-density concrete has been used for such applications [2]. High-density concrete is the most cost-effective material for radiation shielding and has been commonly used for radiotherapy facilities, nuclear reactors, and spent-fuel storage in nuclear power plants [3–5].

In this research work, high-density concrete ($\mu = 4.71$ g/cm^3) using steel balls as aggregates and a normal concrete using recycled aggregates from the debris of the demolished concrete buildings of the earthquake-affected region have been used to evaluate the radiation-shielding property. Two cylindrical model-storage containers for radioactive contaminated-soil samples have been made using these two types of concrete.

Concrete containers with high-density and normal concrete with recycled aggregates for comparison, have been designed for concrete thicknesses of 100–200 mm, respectively, for equivalent radiation- shielding performance and tested for radiation-shielding performance using cesium radioactive contaminated-soil samples as a volume source. Furthermore, in this study, cesium (^{134}Cs and ^{137}Cs) contaminated-soil samples have been used as a volume radioactive source that cannot be simulated with point or line sources from cobalt (^{60}Co), as are usually used in the experimental evaluation of shielding. The radiation-shielding performance of high-density concrete and normal concrete containers against gamma (γ) rays emitted from a soil volume source contaminated by radioactive cesium, have been evaluated through experiments and analysis. In addition to this, the long-term durability of such containers has been taken into consideration and tested for resistance of high-density concrete to freezing and thawing tests.

2   Materials and Mix Proportions

2.1   High-Density Concrete

Table 1 formulates properties and mix proportions of high-density concrete used to produce cylindrical storage containers. Tap water and ordinary Portland cement as

Table 1 Mix proportions and physical properties of materials for high-density concrete

	Water	Cement	Expansive agent	Iron ball and powder	Chemical admixture
Density (g/cm^3)	1.00	3.16	2.94	7.80	-
Weight ratio (%)	4.90	17.1	1.06	76.5	0.44

Table 2 Mix proportions and physical properties of materials for normal concrete using recycled aggregates

	Water	Cement	Fly ash	Fine aggregate (crushed sand)	Coarse aggregate (recycled aggregate)	Chemical admixture
Density (g/cm^3)	1.00	3.15	2.32	2.61	2.40	-
Weight ratio (%)	7.56	17.6	3.78	34.7	36.2	0.26

the binder was used along with a shrinkage-reducing agent (expansive agent) as an additive with a water-binder (W/B) ratio of 27% (water/cement ratio, W/C = 28.7%). Steel balls with a particle size of >0.5 mm was used as aggregate. In order to prevent segregation of steel balls, a low water-cement ratio and high cement content with a combination of a high-range water-reducing agent was used to attain adequate flow ability. The high-density concrete tested had a flow of 220 mm (JIS R 5201) and an air content of 3.0% in its fresh state. The high-density concrete in its fresh state had a unit weight of 4.71 g/cm^3. The 28d average compressive strength of moist-cured (20 °C, 60% RH) Φ100 × 200 mm cylinders was 69.7 N/mm^2.

2.2 Normal Concrete Using Recycled Aggregates

Table 2 shows the materials and mix proportions of the normal concrete, using recycled aggregates from the demolished concrete-building rubble, to investigate this particular application to recycle the huge amount of concrete debris from the large number of damaged and demolished buildings after the Great East Japan Earthquake. This would help to solve another major social problem related to the recycling of debris from the earthquake disaster. Ordinary Portland cement and fly-ash was used as binder, and tap water was used for the mix with a W/B = 35.4%. The recycled coarse aggregate used was in accordance with JIS A 5021, from Miyagi prefecture concrete rubble and pretested for no radioactivity contamination by a Ge detector. Crushed granite sand was used as the fine aggregate. The freshly mixed concrete with recycled aggregates had a slump-flow of 54 cm (JIS A 1150) and an air-content of 5.8%. The 28d average compressive strength of moist-cured (20 °C, 60% RH) Φ100 × 200 mm cylinders was 38.4 N/mm².

3 Design and Casting of Concrete Containers for Storage of Radioactive Contaminated Soil

Figure 1 illustrates the shape and dimensions of the cylindrical-model storage container for high-density concrete. The cylindrical container with a cover-lid had a storage capacity of 0.028 m³ = 28 l of contaminated soil in the centre. The high-density concrete-shield thickness for all sides is t = 100/mm. The shape of the container made by using normal concrete with recycled aggregates was the same and with same storage capacity in the centre except for the concrete-shield thickness, which was t = 200 mm, twice the thickness of high-density concrete. The concrete-shield thickness ratio of high-density and normal concrete of 1:2 had a total weight lower than normal concrete and was designed for an equivalent radiation-shielding performance. Figure 2 and Table 3 gives the shape and dimensions of the concrete-storage containers. Figure 3 shows the containers made from high-density concrete and normal concrete using recycled aggregates.

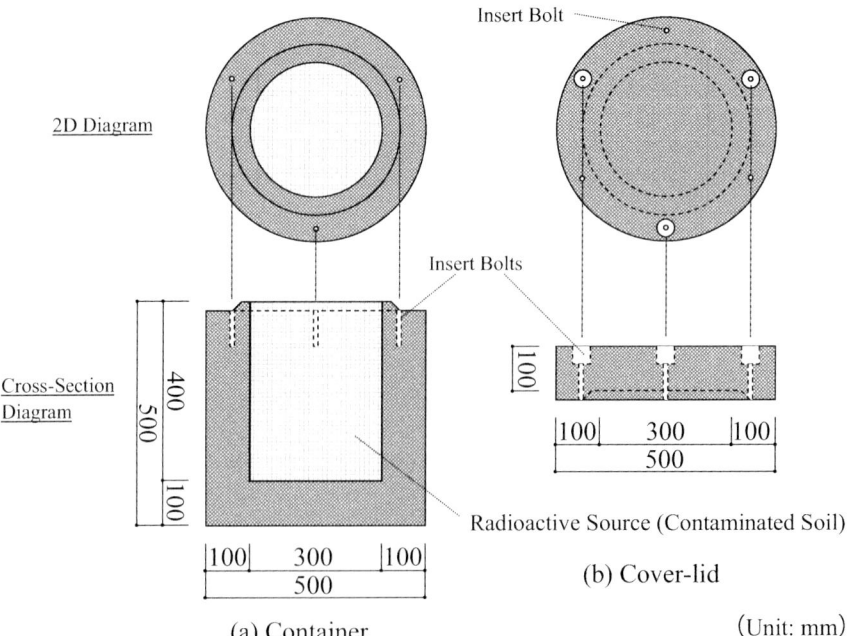

Fig. 1 Details of high-density concrete container and cover

Fig. 2 Shape of container

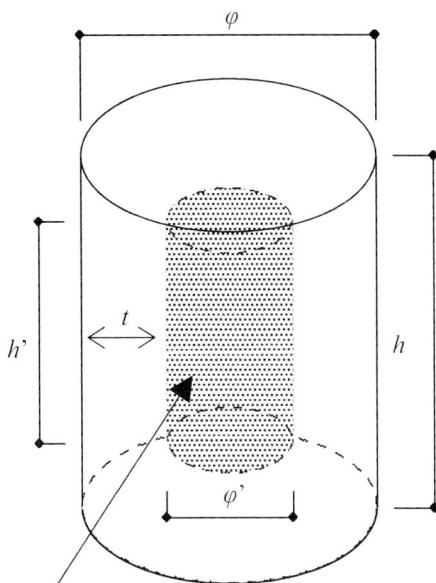

Radioactive Source (Contaminated Soil)

Table 3 Dimensions of storage containers

Identification	High-density concrete (mm)	Recycled concrete (mm)
t	100	200
φ	500	700
h	600	800
φ'	300	
h'	400	

4 Radioactive Contaminated-Soil Samples

The radioactive contaminated-soil samples were taken from four different building sites in the surrounding area in Fukushima, located 70 km from the nuclear-power-station accident site in Dec., 2011. Four samples, each containing 20 kg bags of contaminated soil, were measured for radioactivity using a portable dose meter. In order to attain uniform radioactivity level all the four bags were mixed homogenously and 72 samples of each 50 g were extracted from the mix and tested for radioactivity using a Ge detector. The average radioactivity of 72 samples of contaminated soil for ^{134}Cs and ^{137}Cs were 31.4 ± 4.0 Bq/g and 48.0 ± 6.2 Bq/g, respectively.

Fig. 3 A view of concrete containers

Fig. 4 A view of the
measurement set-up

5 Test Methods for Radioactivity-Shielding Measurements

Figures 2, 4 and 5 show the measurement method and test conditions for
radioactive contaminated- soil samples with and without concrete shielding. The
environmental background radioactivity of the test site was 0.5 μSv/h. The
radioactive contaminated samples were packed in 1-mm thick polyethylene bag in a
cylindrical shape with a diameter of 300 mm and height of 400 mm, and measurements were carried out for the following three conditions:

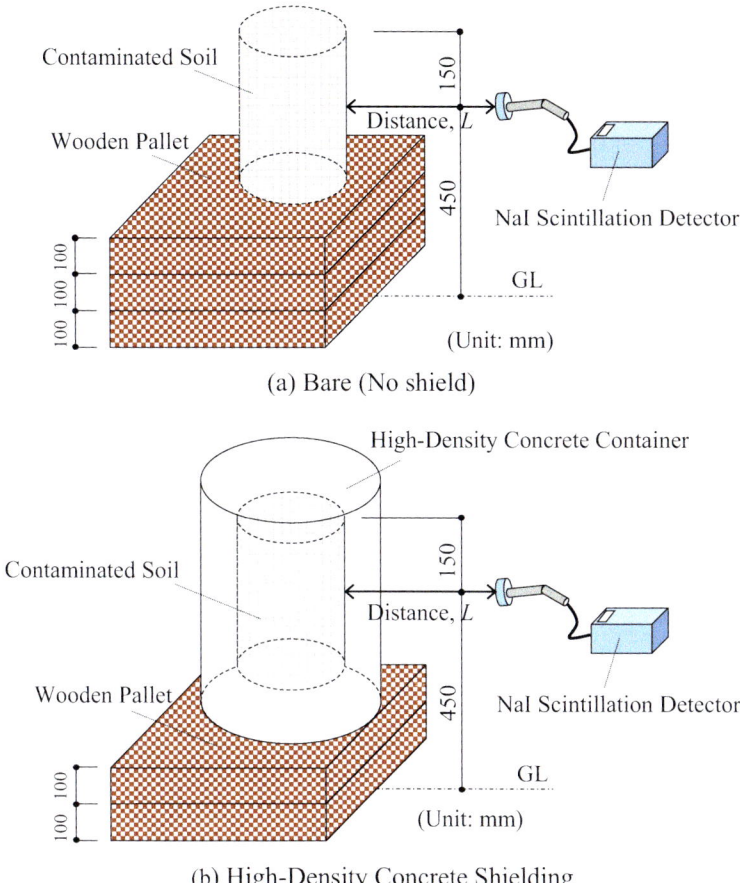

Fig. 5 Measuring methods for radiation shielding

(1) Radioactivity contaminated bare-soil samples in polyethylene bags with no shielding (Bare)
(2) Shielding of radioactive contaminated-soil samples by high-density concrete (H-Con)
(3) Shielding of radioactive contaminated-soil samples by normal concrete with recycled aggregates (N-Con).

All the measurements were carried out in such a position that the centre of soil remained at a ground height of 450 mm as shown in Fig. 5. The measurements for radioactivity shielding by high-density concrete were conducted for three points at a distance of L = 0.1, 0.2 and 0.3 m from the centre point of the sample and for normal concrete with recycled aggregates for two points at a distance of L = 0.2

and 0.3 m since the concrete-shield thickness was 200 mm. Survey Instrument Model 5000 was used for the measurements. Each measurement was carried out for 3 mins and was an average of 6–8 measurements for each point. The influence of the environmental background radiation was deducted from the readings for the shielding effect of the concrete containers.

6 Test Results and Discussion

6.1 Radiation-Shielding Performance of Concrete

Table 4 gives the radiation-shielding performance of high-density concrete and normal concrete with recycled aggregates, in comparison to the measurements for the contaminated bare-soil samples without shielding for the respective measurement distance L. The shielding ratio or attenuation of high-density concrete and normal concrete with recycled aggregates is a ratio of the respective measurement of concrete to the bare contaminated soil sample.

Figure 6 illustrates the measurements of radiation dose rates of the contaminated soil sample for L = 0.1, 0.2 and 0.3 m, respectively. It is clearly evident that the radiation-dose rate is inversely proportional to the distance (L) and shows a drastic decrease with an increase in distance of the measurement point, irrespective of the shielding-concrete thickness or type of concrete. A remarkable attenuation of 94% was observed for high-density concrete which measured 0.18 μSv/h at the distance of L = 0.1 m in comparison to the contaminated bare-soil samples which measured 3.15 μSv/h. Similarly, the measurements of radioactive dosage at a distance of L = 0.2 m for contaminated bare-soil sample, high-density (H-Con) and normal (N-Con) concrete was 1.71 μSv/h, 0.14 μSv/h and 0.15 μSv/h, respectively. Furthermore, at a distance of L = 0.3 m for bare-soil sample, high-density and normal concrete was 1.07 μSv/h, 0.07 μSv/h, and 0.09 μSv/h, respectively, showing a remarkable decrease in radioactivity dosage. The shielding performance by containers made with high-density concrete with a thickness of 100 mm and normal concrete, using recycled aggregates with a thickness of 200 mm, showed more than 90% attenuation or reduction of radioactivity dosage. The radioactivity attenuation

Table 4 Radiation shielding performance by high-density concrete and normal concrete

Measurement distance, L (m)	Bare soil (μSv/h)	H-Con shielding (μSv/h)	N-Con shielding (μSv/h)	Shielding or attenuation by H-Con (%)	Shielding or attenuation by N-Con (%)	H-Con/N-Con
0.1	3.15	0.18	-	94.3	-	-
0.2	1.71	0.14	0.15	91.8	91.2	1.01
0.3	1.07	0.07	0.09	93.5	91.6	1.02

Fig. 6 Effect of
concrete-shield thickness or
measurement distance on
radioactive dose rates

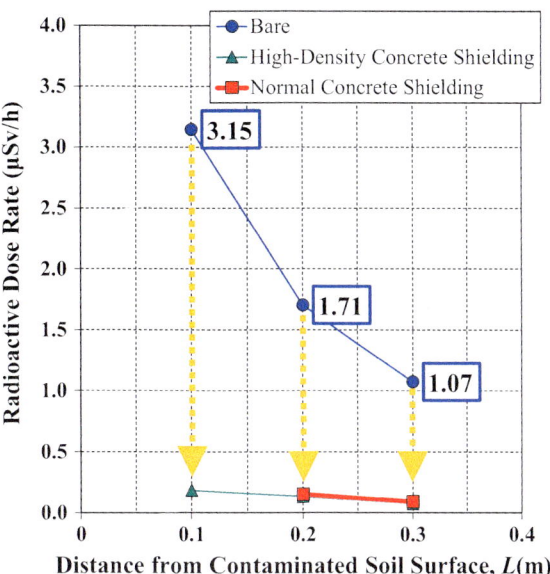

rates by concrete containers made with high-density concrete (t = 100 mm) in
comparison to normal concrete with recycled aggregates (t = 200 mm) was 1.01
times at L = 0.2 m and 1.02 times at L = 0.3 m showing equivalent or higher
shielding performance.

6.2 Analysis and Simulation of Experimental Results

Analysis of radiation shielding was performed using Monte Carlo N-Particle
Transport Code System (MCNP42C) using photo library (MCPLIB02) with "radiation dose conversion coefficients fpr radiation shielding calculations based on the
AESJ-SC-R002:2010 and the results were compared to those of experimental
measurements [6, 7]. The measured densities of the contaminated-soil samples
$\mu = 1.25$ g/cm^3, high-density concrete $\mu = 4.57$ g/cm^3 and normal concrete with
recycled aggregates $\mu = 2.14$ g/cm^3 was used for calculations. For the simulations,
respective concrete was assumed to be uniform and isotropic. The
contaminated-soil sample was also assumed to be uniform and the atomic distribution of ^{134}Cs and ^{137}Cs as the isotropic photon- volume source. As the shielding
from gamma (γ) rays is strongly influenced by the density and concrete-shielding
thickness, the equivalent number of atoms for each material was assumed for
analysis purpose to constitute the same density as measured.

Figure 7 shows the experimental and analytical results of attenuation curves for
radiation-dose rates for high-density concrete, normal concrete and contaminated
bare-soil samples, as a function of the measurement distance (L) or concrete-shield

Fig. 7 Experimental and
analytical results for concrete
shield thickness versus
radioactive dose rates

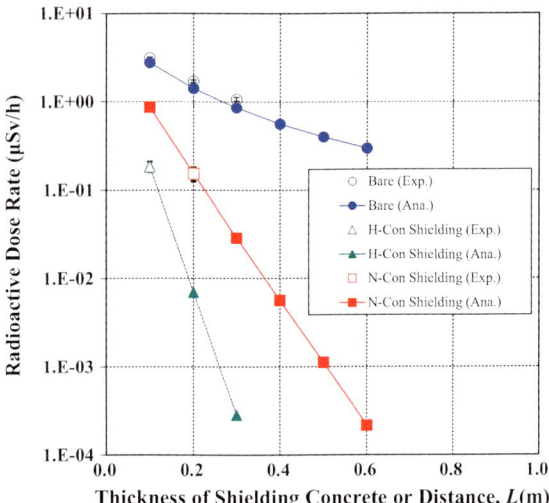

thickness (t). The analysis results were simulated for two types of concretes with various shield thicknesses. The experimental results were only for a single concrete-shield thickness (0.1 m for H-Con and 0.2 m for N-Con) and fit well on the analytical curves for high-density concrete and normal concrete with recycled aggregates. The results of the analysis matched well with the experimental results of the contaminated bare sample and the variation of data was below 15%. The results of the MCNP analysis for simulated conditions are in good agreement with the experimentally measured results. In addition, high-density concrete shows the same shielding performance as the normal concrete with twice the thickness of concrete shielding, as compared to the high-density concrete.

6.3   Design for Optimization of Concrete-Shield Thickness

Figure 8 illustrates the concrete-shield thickness vs. ratio of dose transmission. The curves represent the intensity of radiation transmission through high-density concrete ($\mu = 4.57$ g/cm^3) and normal concrete ($\mu = 2.14$ g/cm^3). The calculated values from the analysis were below the intensity of radiation transmission curves, and the difference is higher with increased shield thickness. These differences in values are due to the use of point-radiation sources for analysis, which is different from the volume-radiation source from the contaminated soil. Therefore, for the optimization of design-shield thickness of concrete for containers, analysis methods using volume-radiation source need to be taken into consideration [7].

Fig. 8 Analytical results for shield thickness versus radioactive dose transmission

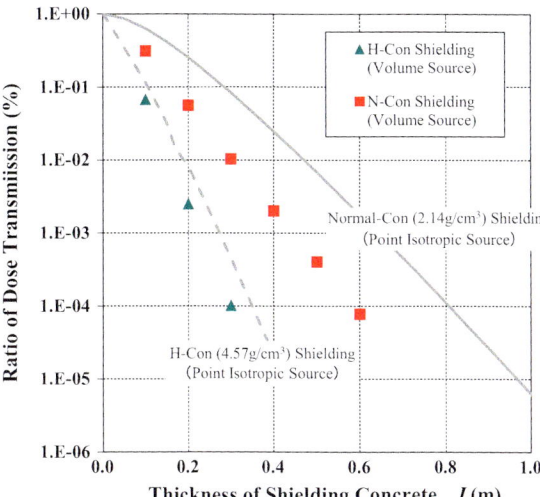

Fig. 9 Number of freeze-thaw cycles versus relative dynamic modulus of elasticity of concretes

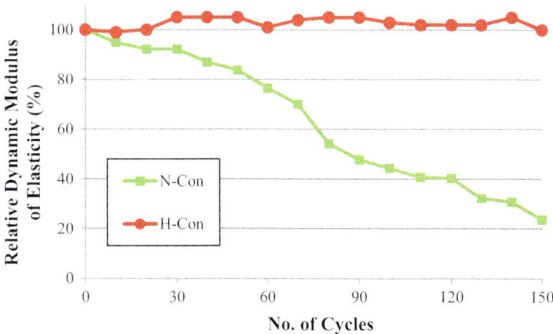

6.4 Freezing and Thawing Resistance of High-Density Concrete

The freezing and thawing resistance of high-density concrete is an important factor for using it as storage containers for radioactive contaminated wet-soil samples. Figure 9 gives the relative dynamic modulus of elasticity of high-density concrete and normal concrete for 0–150 freeze-thaw cycles. A dramatic loss of relative dynamic modulus of elasticity for normal concrete is observed after 50–60 cycles in comparison to high-density concrete. At 100 freeze-thaw cycles, the relative modulus of elasticity of normal concrete was hardly 40% of the initial stage in comparison to high-density concrete, which remained unchanged, showing a superior freeze-thaw resistance. Figure 10 shows the weight change of high-density and normal concrete specimens subjected to freezing and thawing cycles. The weight change of specimens is directly proportional to the deterioration of relative

Fig. 10 Effect of freeze-thaw on weight change of concrete specimens

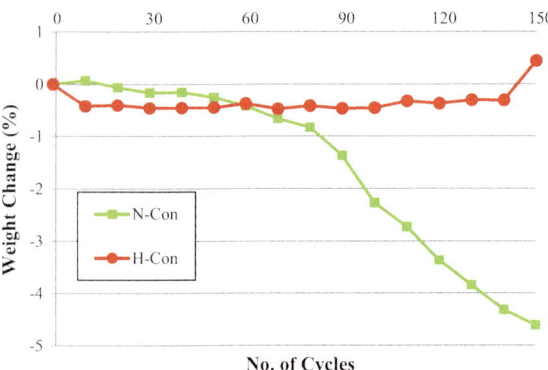

dynamic modulus of elasticity of the specimens. An abrupt weight loss due to scaling is observed for normal concrete specimens after 50 cycles and more cycles, whereas the weight change of high-density concrete does not show a distinct loss in weight of the specimens.

7   Conclusions

Two types of model cylindrical containers for storage of radioactive [134]Cs and ^{137}Cs contaminated soil in Fukushima were made, using high-density concrete ($\mu = 4.57$ g/cm^3) and normal concrete with recycled aggregates ($\mu = 2.14$ g/cm^3) and were evaluated for shielding performance. From the results of the experiments, it was demonstrated that the cylindrical container using high-density concrete of 100 mm shield thickness had attenuation as high as 90% for radiation-dosage rates emitted from radioactive cesium in contaminated soil. Good agreement was observed between the experimental and analytical values using Monte Carlo MCNP4C2 simulation for shielding performance. Also, the freezing and thawing durability of high-density concrete was found to be superior to normal concrete using recycled aggregates. Therefore, the proposed high-density concrete with adequate strength and durability is suitable for shielding the radioactive contaminated soil of Fukushima.

Acknowledgements This research was partially supported by Grant-in-Aid for Young Scientists (B) No. 25820265 provided by Japan Society for the Promotion of Science (JSPS).

References

1. The Society for Remediation of Radioactive Contamination in Environment (2012) Abstracts of 1st research presentations on remediation of radioactive contamination in environment (In Japanese)
2. Cheyrezy HM (1996) Development of HPC in France: recent achievement and future trends. Am Concrete Inst Special Publ 159:145–158
3. Davidovits Joseph (1994) Recent progresses in concretes for nuclear waste and uranium waste containment. Am Concrete Inst Concrete Int 16(12):53–58
4. Peng Y-C, Hwang C-L (2011) Development of high performance and high strength heavy concrete for radiation shielding structures. Int J Miner Metal Mater 18(1):89–93
5. Mortazavi SMJ, Mosleh-Shirazi MA, Roshan-Shomal P, Raadpey N, Baradaran-Ghahfarokhi M (2010) High-performance heavy concrete as a multi-purpose shield. Radiat Prot Dosimetry 142(2–4):120–124
6. Los Alamos National Laboratory (2003) MCNP4C Monte Carlo N-Particle Transport Code System, CCC-700
7. Suzuki Y, Kimura K, Fujikura Y, Lee Y, Pareek S, Araki Y (2013) Concrete Res Technol 24(2):43–52 (In Japanese)

Ageing Behaviour of Neutron/Irradiated Eurofer97

I. Carvalho, A. Fedorov, M. Kolluri, N. Luzginova, H. Schut and J. Sietsma

Abstract Eurofer97 is a candidate structural material for nuclear fusion reactors. To better understand the ageing effects due to radiation during long-term use in real fusion conditions, Eurofer97 was neutron irradiated at the High Flux Reactor in The Netherlands. TEM images of post-irradiation Eurofer97 reveal a high density of irradiation-induced dislocation loops, fine precipitates and agglomeration of point defects. Microscopy results are related to post-irradiation tensile mechanical tests done at room temperature and at 300 °C. An increase of yield and ultimate tensile strength combined with a decrease of total elongation is observed in both tests and correlated with the presence of radiation damage.

Keywords Eurofer chromium steel · Radiation damage · Nuclear fusion
High flux reactor · Mechanical properties

1 Introduction

As the world population grows and with it the demand for energy generation, new energy- supply options are to be considered. The general acceptance of nuclear technology depends on operational safety, sustainability, waste management and

I. Carvalho
Materials innovation institute (M2i) Delft, Delft, The Netherlands

I. Carvalho · A. Fedorov · M. Kolluri · N. Luzginova
Nuclear Research and Consultancy Group (NRG), Petten, The Netherlands

I. Carvalho · H. Schut
Faculty of Applied Sciences, Delft University of Technology (TUDelft), Delft, The Netherlands

J. Sietsma (✉)
Faculty of Mechanical, Maritime and Materials Engineering, Delft University of Technology (TUDelft), Delft, The Netherlands

© Springer International Publishing AG 2018
K. van Breugel et al. (eds.), *The Ageing of Materials and Structures*,
https://doi.org/10.1007/978-3-319-70194-3_9

111

economical gains [1]. For the existing fission nuclear-power plants it is crucial to monitor the degradation of the reactor components and the materials properties. Alternatively, fusion power plants are being researched and developed. Fusion energy offers a long-term secure source of energy, with the advantage of no production of greenhouse gases and no long-lived radioactive waste.

Eurofer97 is a Reduced Activation Ferritic/Martensitic (RAFM) steel used as a reference structural material for future fusion reactors. Eurofer97 is known to possess a high resistance against swelling caused by gaseous transmutation products (hydrogen, helium), has attractive mechanical properties and shows reduced activation behaviour in a fusion neutron spectrum [2]. The ageing effects of neutron irradiation and the accumulation of defects in this material are extensively studied as Eurofer97 will be used for constructing test blanket modules in the ITER fusion reactor [2, 3]. As a consequence of the neutron irradiation, displacement damage will be created, lead to changes in the microstructure and, ultimately, to alterations in the material's mechanical properties [2, 3]. Understanding the microstructural ageing of RAFM steels under neutron-irradiation conditions and its effect on the mechanical properties is of critical importance for the design and application of Eurofer97 at ITER and future fusion-power plants.

The chemical composition of this steel contains Fe, Cr, W, V, Ti, Ta and C, which are low activation elements that allow recycling of the waste in 100 years' time [2]. As a structural material, Eurofer97 will not only be exposed to a 14 MeV neutron environment but also to thermo-mechanical loading. Both conditions cause microscopic alterations. Although Eurofer97 has a low sensitivity to radiation-induced swelling and helium embrittlement under neutron irradiation [2], a detailed understanding of the effects of irradiation and temperature is crucial for a correct design and application of Eurofer97 in a fusion reactor. Analysis of the mechanical properties of Eurofer97 has been previously reported [4, 5], but a correlation with the type of defects found in the material after irradiation has not yet been clearly established. This work is among the first that aim to establish the relationship between radiation damage and deterioration of mechanical properties.

In order to mimic in a short period of time the ageing effects of the long-term use of Eurofer97 in a fusion reactor, this steel was neutron irradiated at the High Flux Reactor (HFR) in Petten, The Netherlands. Irradiation programs with temperatures of 60 and 300 °C and neutron doses of 2.5 and 10 displacements per atom (dpa) [4, 5] were completed. This paper focusses on the 10 dpa and 300 °C irradiation. To comprehend the consequences of the irradiation damage on the mechanical properties of the Eurofer97, Transmission Electron Microscopy (TEM) investigations are correlated with post-irradiation mechanical tensile tests.

2   Experiments

2.1   The Material

The European Union batches of Eurofer97 were produced in Böhler, Austria. The nominal composition of Eurofer97 is Fe-9Cr-1 W-0.2 V-0.1Ta-0.1C (wt%). After fabrication the material was subjected to austenitization at 980 °C for 30 min, followed by annealing at 760 °C for 90 min to achieve a tempered martensitic structure [6]. Both steps were followed by air cooling.

2.2   Irradiation Programme at HFR

Eurofer97 material was irradiated in multiple campaigns at HFR [5]. The irradiation programme of the piece of Eurofer97 piece that was investigated by TEM is designated 'in-SodiUm steel Mixed specimens irradiation 04' (SUMO-04). In this programme, a set of various different steels, already cut to the correct specimen shape to perform mechanical post-irradiation examination, were irradiated with 10 dpa at a temperature of 300 °C. The calculated helium content is 12.6 appm.

The specimens are placed in irradiation rigs, which are filled with sodium to ensure good heat conductivity. The irradiation temperature is determined by the balance between the gamma heating and the heat dissipation via the gas gaps introduced in the sample holder [4]. The temperature of the irradiation is controlled with 20 thermocouples mounted on the specimens. As for the neutron monitoring, 13 detectors are placed in key positions within the irradiation rig. The uncertainty of the irradiation dose in terms of dpa is 14% [4].

2.3   Sample Preparation

TEM discs were cut from broken pieces of a fracture-toughness specimen that was post-irradiation mechanically tested at room temperature, in a region away from the fracture zone. Discs with a diameter of 3 mm and a thickness of approximately 100 μm were manufactured by a sequence of grinding and polishing steps in hot cells. The final thinning was done by electro-polishing with a solution of 135 ml of acetic acid and 15 ml of per-chloric acid.

2.4   TEM Examination

The TEM investigation reported here was done at NRG using a JEOL 1200ex STEM/TEM microscope with an accelerating voltage of 120 kV.

3   Results

3.1   Microstructure of Unirradiated Eurofer97

TEM images of unirradiated Eurofer97 are shown in Fig. 1 [7]. On the left, the overview of the sample reveals the typical lath martensitic structure expected for Eurofer97 [8]. According to specifications [2], the Eurofer97 used in this work has a grain size in the range 9–23 μm. Coarse, spheroidally shaped particles are present mostly along the grain boundaries and have a size of ~100 nm. Although at a lower density, platelet particles are observed with sizes in the range of 100–200 nm [7]. The composition of the particles was not studied, but the observed big (~100 nm) spheroidally shaped particles can be recognised as $M_{23}C_6$ or MX-type phases (where M is a Fe, Cr or W and X is Ta or V), based on the literature [8, 9].

3.2   Microstructure of 10 dpa, 300 °C Neutron-Irradiated Eurofer97

Figure 2 shows a low-magnification TEM image of 10 dpa, 300 °C neutron-irradiated Eurofer97. In the figure, the pre-irradiation lath structure [7, 8] is observed. Precipitates are observed mostly around the grain boundaries but also

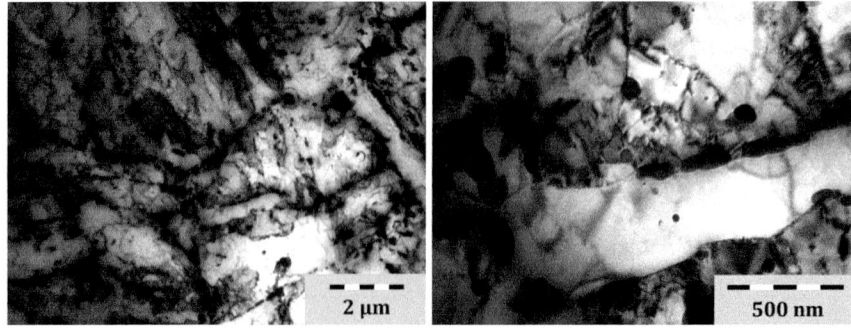

Fig. 1 Unirradiated Eurofer97. On the left is an overview of the lath microstructure expected for Eurofer97. On the right, a detailed image reveals spheroidally shaped particles present close to grain boundaries [7]

Fig. 2 Overview of the microstructure of 10 dpa 300 °C neutron-irradiated Eurofer97 irradiated at HFR. The pre-irradiation lath can be observed. Radiation damage is spread throughout the sample

Fig. 3 Detail of the microstructure of 10 dpa 300 °C neutron-irradiated Eurofer97 showing the irradiation induced defects. Black dots are marked by arrows, and dislocation loops are encircled

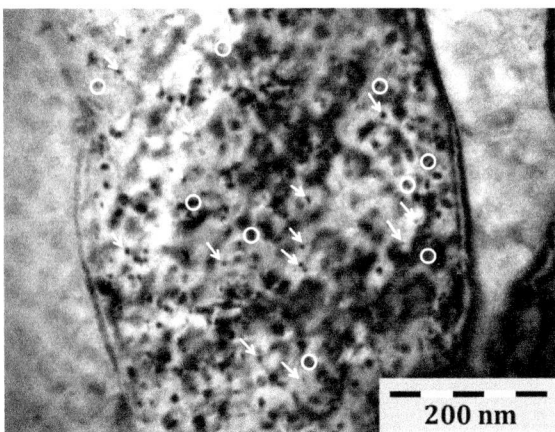

inside the grains. The composition of the precipitates was not investigated but, as noted for unirradiated Eurofer97, these defects are expected to be $M_{23}C_6$ or MX-type precipitates [9]. Furthermore, other authors identified the formation of small α'- (chromium-rich precipitates) and M_6C-phases inside the grain as a consequence of irradiation-enhanced diffusion [10, 11]. The dimensions of the precipitates range between 20 and 200 nm. Radiation damage, characterized by a high density of uniformly distributed "black dots" (identified as an agglomeration of point defects or fine precipitates [1]), is found throughout the material.

In Fig. 3, an image of a Eurofer97 grain damaged by ageing from neutrons is shown. Two types of radiation damage are distinguishable: black dots and dislocation loops. The black dots reach dimensions of approximately 10 nm and are dispersed uniformly in the grain. The dislocation loops reach dimensions of 15 nm and are also located throughout the grain.

4   Discussion

Comparing the TEM investigations of pre- and post-neutron irradiation Eurofer97, significant microstructural changes are observed to be induced in the latter condition. These changes appear as dislocation loops and "black dots", definition given to irradiation-induced defects, such as small Frank loops or defect clusters observed with TEM, but not clearly identifiable [1].

Post-irradiation mechanical tensile tests of Eurofer97 are published elsewhere [4, 5]. The stress-strain curves are shown in Fig. 4. Three curves are plotted: one for Eurofer97 before irradiation, used as a reference curve, and two regarding post-irradiation tensile tests done at different temperatures—one at room temperature (RT) and the other at 300 °C, the same temperature as used for the neutron irradiation. In the figure, the curves are plotted starting at the onset of the plastic region. The elastic region of the curves is not shown as it was affected by the tensile testing-machine compliance used for the measurements and is not relevant for this analysis. For the three curves, the same strain rate and specimen cross section were used: 5×10^{-4} s^{-1} and 12.57 mm^2, respectively. The yield strength (YS) and the ultimate tensile strength (UTS) values of unirradiated Eurofer are 550 and 692 MPa, respectively. After irradiation, the YS and UTS for each tensile-test temperature have the same value, i.e. 1066 MPa for the RT test and 883 MPa for the 300 °C test.

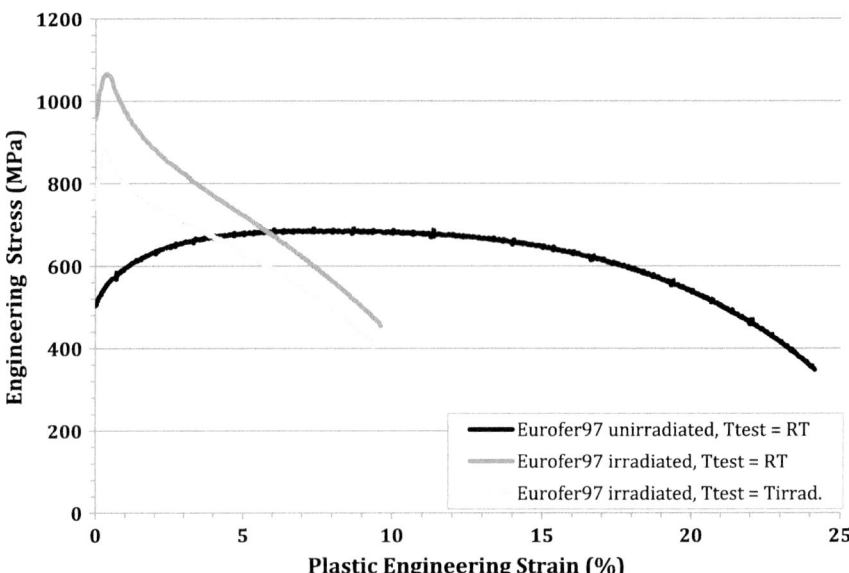

Fig. 4 Plastic region stress-strain curves of unirradiated and neutron-irradiated 10 dpa, 300 °C Eurofer97. Test temperatures are indicated in the legend. The curves are plotted from the onset of the plastic region

The reference curve of Eurofer97 shows a long plastic region before necking. In the irradiated samples' curves, the uniform elongation is practically zero, and the total elongation has significantly decreased. The black dots and dislocation loops observed in Fig. 3 are envisaged to be responsible for the increase in YS and UTS after irradiation in both conditions. In the tensile test performed at RT, the black dots and precipitates act as obstacles for dislocation movement. In the test done at 300 °C, the same temperature used for irradiation, the YS is lower because of thermally-activated dislocation gliding promoted by the increased temperature. Annihilation of defects is unlikely to be related to the lower YS and UTS of this test curve as the test did not take longer than 200 s at 300 °C.

5 Conclusions

A comparison of Eurofer97 TEM investigations before and after 10 dpa and 300 °C neutron irradiation reveals that a significant amount of lattice damage is created by the neutron exposure. For both conditions, precipitates are observed mainly along the grain boundaries. The radiation damage induced by neutrons appears as black dots and dislocation loops. The black dots are very small and unresolved dislocation loops, defect clusters or small α'- and M_6C-phases [9, 10]. The resolved dislocation loops can reach sizes of 15 nm.

The TEM images are correlated to post-irradiation tensile tests published elsewhere [4, 5]. In comparison to values of reference unirradiated Eurofer97, an increase in yield strength and ultimate tensile strength after neutron irradiation is observed for the tensile tests performed at room temperature and at the irradiation temperature (300 °C). Both irradiated Eurofer97 samples show a significant decrease in uniform and total elongation. The lattice damage caused by irradiation and observed with TEM is held responsible for the irradiation-induced hardening. Dislocation gliding is promoted at the 300 °C tensile-test temperature, manifested as a decrease in YS and UTS in comparison to the values obtained for the neutron-irradiated specimen tested at room temperature.

Acknowledgements This research was carried out under project number M74.5.10393 in the framework of the Research Programme of the Materials Innovation Institute M2i (http://www.m2i. nl). The authors thank F. v.d. Berg for his support in the TEM measurements and M. Klimenkov for discussions concerning the TEM images.

References

1. Koning R (2012) comprehensive nuclear materials, 1st edn. Elsevier, United Kingdom
2. Lindau R, Moslang A, Rieth M, Klimiankou M, Materna-Morris E, Alamo A, Tavassoli F, Cayron C, Lancha AM, Fernandez P, Baluc N, Schaublin R, Diegele E, Filacchioni G, Rensman JW, vd Schaaf B, Lucon E, Dietz W (2005) Present development status of

EUROFER and ODS-EUROFER for application in blanket concepts. Fusion Eng Des 75–79:989–996

3. vd Schaaf B, Tavassoli F, Fazio C, Rigal E, Diegele E, Lindau R, LeMarois G (2003) The development of EUROFER reduced activation steel. Fusion Eng Des 69:197–203
4. Rensman J (2005) NRG irradiation testing: report on 300 and 60 °C irradiated RAFM Steels, NRG 20023/05.68497/P, Petten
5. Luzginova N, Rensman J, Jong M, t Pierick P, Bakker T, Nolles H (2014) Overview of 10 years of irradiation projects on Eurofer97 steel at high flux reactor in Petten. J Nucl Mater 455:21–25
6. Rieth M, Schirra M, Falkenstein A, Graf P, Heger S, Kempe H, Lindau R, Zimmermman H (2003) Eurofer 97: tensile, charpy, creep and structural tests. FZKA, report 6911
7. Kolluri M, Edmondson PD, Luzginova N, vd Berg F (2014) A structure-property correlation study of neutron irradiation induced damage in EU batch of ODS Eurofer97 steel. Mater Sci Eng A 597:111–116
8. Klimenkov M, Lindau R, Materna-Morris E, Moslang A (2012) TEM characterization of precipitates in EUROFER 97. Prog Nucl Energy 57:8–13
9. Fernandez P, Lancha AM, Lapena J, Serrano M, Hernandez-Mayoral M (2002) Metallurgical properties of reduced activation martensitic steel Eurofer'97 in the as-received condition and after thermal ageing. J Nucl Mater 307:495–499
10. Gaganidze E, Petersen C, Materna-Morris E, Dethloff C, Weiss OJ, Aktaa J, Povstyanko A, Fedoseev A, Makaroc O, Prokhorov V (2011) Mechanical properties and TEM examination of RAFM steels irradiated up to 70 dpa in BOR-60. J Nucl Mater 417:93–98
11. Materna-Morris E, Moslang A, Rolli R, Schneider H (2009) Effect of helium on tensile properties and microstructure in 9% Cr/WVTa/steel after neutron irradiation up to 15 dpa between 250 and 450 °C. J Nucl Mater 386–388:422–425

Ageing Management for Extended Storage of Spent Nuclear Fuel

Y. Y. Liu

Abstract This chapter summarises issues related to managing aging effects on the structures, systems and components that are associated with the confinement boundary of storage casks and canisters, with particular emphasis on the ageing-management programme for monitoring the structural and functional integrity of the internals in the canister or cask, including the spent-fuel assemblies. Specific examples, such as potential embrittlement of high-burnup cladding due to hydride reorientation during vacuum drying, transfer and early storage, thermal modeling, and development of advanced surveillance technology, are used to highlight the key elements common within the aging-management programmes that emphasise prevention, mitigation, condition and/or performance monitoring.

Keywords Ageing management · Spent-nuclear fuel · Storage Transportation · Monitoring · Inspection

1 Introduction

The reactor core of a nuclear power plant in the USA consists of 100–1,000 fuel assemblies. These fuel assemblies are typically replaced after four to six years. The irradiated nuclear-fuel assemblies, commonly referred to as "used" or "spent"-fuel assemblies, are highly radioactive and thermally hot. They are initially stored in an on-site, steel-lined storage pool to help shield the radiation and cool the fuel. The spent-fuel pools were intended to serve as temporary storage facilities until the used-fuel assemblies could be safely transferred to a permanent storage repository or a reprocessing facility. However, starting in the 1970s, interest in the commercial reprocessing of spent fuel diminished in the USA, and progress toward a permanent storage repository continued to fall behind schedule, forcing nuclear power plant operators to transfer spent fuel from spent-fuel pools or wet storage into dry-cask

Y. Y. Liu (✉)
Argonne National Laboratory, Argonne, IL, USA
e-mail: yyliu@anl.gov

© Springer International Publishing AG 2018
K. van Breugel et al. (eds.), *The Ageing of Materials and Structures*,
https://doi.org/10.1007/978-3-319-70194-3_10

storage systems (DCSSs). Furthermore, delay and uncertainty in the ultimate disposition of used fuel in the US raises the prospect of extended long-term storage and deferred transportation of spent fuel at operating and decommissioned nuclear power plant sites. The situation is similar worldwide in other countries with civilian nuclear-power programmes.

Under US federal regulations contained in Title 10 of the Code of Federal Regulations (CFR) 72.42, the initial license term for an Independent Spent Fuel Storage Installation (ISFSI) must not exceed 40 years from the date of issuance. Licenses may be renewed by the U.S. Nuclear Regulatory Commission (NRC) at the expiration of the license term upon application by the licensee for a period not to exceed 40 years. Applications for ISFSI license renewals must include (1) time-limited ageing analyses (TLAAs) that demonstrate that structures, systems and components (SSCs) important to safety will continue to perform their intended function for the requested period of extended operation and (2) a description of the ageing-management program (AMP) for the management of issues associated with ageing that could adversely affect SSCs important to safety. This chapter summarises issues related to managing the aging effects on the SSCs associated with the confinement boundary of bolted- and welded-closure storage casks and canisters, with particular emphasis on a generic AMP, called M5, for monitoring the structural and functional integrity of the internals in the canister or cask, including the spent-fuel assemblies. Other generic AMPs and TLAAs recommended for managing the ageing effects of SSCs in DCSSs/ISFSIs can be found in a report [1], which was prepared by Argonne National Laboratory (Argonne) researchers for the U.S. Department of Energy (DOE), following an approach similar to that of the "Generic Aging Lessons Learned" report [2] for the ageing management and license renewal of nuclear power plants. Potential cladding embrittlement due to hydride reorientation during vacuum drying, transfer and early storage, thermal modeling and development of advanced surveillance technology are used as examples to highlight key elements common within the AMPs that emphasize prevention, mitigation, condition and/or performance monitoring. These examples are largely drawn from Argonne's research and development (R&D) programs conducted since 2010 for DOE's Used Fuel Disposition Campaign on extended storage and transportation of spent fuel.

Table 1 summarises the DCSSs presently in common use in the USA. The DCSS designs are of two general types, namely: (1) self-contained shielded metallic casks without an overpack and (2) metallic canisters with a separate overpack to provide radiation shielding and physical protection.

All of the self-contained cask designs listed in Table 1 incorporate bolted-top enclosures with O-ring seals, whereas the canister plus overpack configurations utilize a welded-top closure. Figure 1 shows cutaway views of vertically- and horizontally-oriented dry-storage casks with a bolted closure lid (Fig. 1a) and welded canisters with concrete overpacks (Fig. 1b and c). After fuel loading in the spent-fuel pool and vacuum drying, the cavity of the cask/canister is filled with high-pressure helium before closure to provide an inerting environment during long-term storage.

Table 1 Summary of dry-cask storage systems in common use in the USA [1]

Vendor	System	Cask/Canister	Type[a]	Closure
Energy Solutions	Fuel Solutions	VSC-24, W150	C/O	Welded
General Nuclear Systems, Inc.	CASTOR	V/21, X/33	Cask	Bolted
Holtec International	HI-STAR 100 HI-STORM 100	MPC-68, MPC-80 MPC-24, MPC-32, MPC-68	C/O C/O	Welded Welded
NAC International, Inc.	S/T MPC UMS MAGNASTOR	NAC-I28 MPC-26, MPC-36 UMS-24 MAGNASTOR	Cask C/O C/O C/O	Bolted Welded Welded Welded
Transnuclear, Inc.	NUHOMS TN Metal Casks	52B, 61BT, 61BTH, 7P, 24P, 24PHB, 24PT, 24PTH, 24PT1, 32P, 32PT, 32PTH, 12T TN-24, TN-32, TN-40, TN-68	C/O Cask	Welded Bolted
Westinghouse	MC-10	MC-10	Cask	Bolted

[a]C/O: metallic welded canister with overpack; Cask: self-contained metallic cask with bolted lid

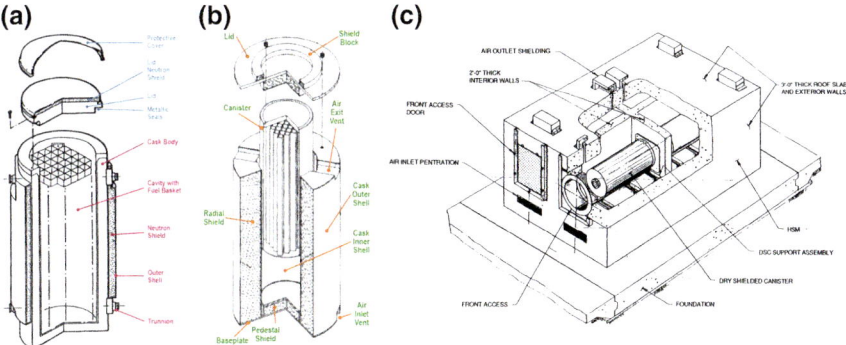

Fig. 1 Cutaway views of vertical and horizontal dry-storage casks with bolted-closure lids (**a**) and welded canisters (**b**) and (**c**), each containing an internal basket for multiple spent-fuel assemblies

2   Program Description of AMP M5 and Interfaces

The objective of AMP M5 in the Argonne report [1] is to manage ageing effects on the structural and functional integrity of the used-fuel assemblies and internals of storage canisters/casks to ensure spent-fuel integrity, fuel retrievability after long-term storage and safe transportation of the used fuel for reprocessing or disposal. The degradation effects to be managed include corrosion, creep, distortion, cracking and peeling of laminates. The program monitors the ageing degradation of the heat transfer, radiation shielding, criticality control, confinement boundary and structural-support functions of the storage canister/cask internals caused by extended exposure to high temperatures and radiation. The programme focuses specifically on high-burnup fuel (HBF) assemblies (i.e. fuel assemblies with burnups exceeding 45 GWd/MTU). The programme consists of: (1) a site-specific assessment of the canister/cask designs to verify compliance with the applicable NRC Interim Staff Guidance (ISG) documents; (2) detailed thermal and radiation analyses of the canister/cask designs used at the site to select cask/canister surface locations that are likely to be most sensitive to changes in temperature and radiation due to potential degradation of the used-fuel cladding and storage-basket assembly or from helium leakage; and (3) development of a monitoring program for continuous measurement of radiation and temperature at these locations to establish data trends under normal operating conditions. Any significant deviation from the established data trend is evaluated for potential degradation of the functional and structural integrity of the canister/cask internals.

The degradation of the neutron-absorbing materials inside the canister/cask is addressed by a TLAA described in Chap. 3 of the report [1], along with other generic TLAAs. The ageing effects of a breach in the canister/cask confinement boundary and helium leakage are managed by either AMP M3, "Welded Canister Seal and Leakage Monitoring Program," or AMP M4, "Bolted Cask Seal and Leakage Monitoring Program," and the ageing effects on the external surface of the canister/cask are managed by AMP M1, "External Surfaces Monitoring of Mechanical Components," and described in Chap. 4 of the report [1]. AMP M5, as stated earlier, manages the structural and functional integrity of storage canister/cask internals to ensure retrievability of the used-fuel assemblies without releasing radioactive material. Spent-fuel assemblies are likely to be retrievable if the fuel rods as well as the assemblies are not warped, the cladding is intact and the confinement boundary has not been breached so as to avoid loss of the inerting environment.

Regulation 10 CFR 72.44 requires that the technical specifications of the DCSS designs include functional and operating limits, monitoring instruments and limiting control settings to protect the integrity of the stored fuel assemblies and to guard against the uncontrolled release of radioactive material. The limiting conditions are the lowest functional capability or performance levels of equipment required for safe operation. Furthermore, 10 CFR 72.44 (c) also requires that the technical specifications of DCSS designs include surveillance requirements for inspection, monitoring, testing and calibration activities to confirm that operation of the canister/cask is within the required functional and operating limits and that the limiting conditions for safe storage are met. One limiting condition for operation specified in the technical specifications is the cavity pressure in vacuum drying, and the dryness is demonstrated by evacuating the cavity to a relatively high vacuum (e.g. ≤ 3 torr [133.3 Pa]) and verifying that a vacuum is maintained over a specified time period (e.g. ≥ 30 min).

2.1 Interim Staff Guidance for AMP M5

The applicable NRC ISG documents referenced for AMP M5 include the following: ISG-2, "Fuel Retrievability"; ISG-11, "Cladding Considerations for the Transportation and Storage of Spent Fuel"; ISG-15, "Materials Evaluation"; ISG-18, "The Design and Testing of Lid Welds on Austenitic Stainless Steel Canisters as the Confinement Boundary for Spent Fuel Storage"; ISG-24, "The Use of Demonstration Program as Confirmation of Integrity for Continued Storage of High Burnup Fuel Beyond 20 Years"; and ISG-25, "Pressure and Helium Leakage Testing of the Confinement Boundary of Spent Fuel Dry Storage Systems." In a recent paper on ageing management [3], a senior staff member from the NRC stated the following:

Guidance, in a regulatory sense, is a recommendation of a method to meet a regulation or conduct an evaluation. Guidance is based on knowledge at the time the guidance was developed and may change as new information is obtained.

Two examples of guidance issued by NRC staff were used to illustrate how new technical information and changes in circumstances can affect the use of guidance documents [3]. The first example involves ISG-18 and -25 that could be combined to imply that, if welds are made to meet certain criteria, then confinement is not an issue. However, these guidance documents were developed at the time when canister stress-corrosion cracking in a marine environment was not identified as a potential ageing effect and degradation mechanism for the canister closure welds. Both NRC and industry personnel are conducting studies to determine whether the initial hypothesis is correct.

The second example is ISG-11, Rev. 3, published in November 2003. This ISG indicates that, if the cladding temperature is held below 400 °C during operation, no cladding degradation will occur, and this finding has been validated for low-burnup fuel and is accepted by NRC as proof of behaviour for extended periods. For HBF, however, no such confirmation exists. DOE is sponsoring a High Burnup Dry Storage Cask Research and Development Project [4], and a team led by the Electric Power Research Institute (EPRI) is to collect data from a dry cask containing HBF over a prolonged period. A TN-32 cask with a bolted-closure lid, shown in Fig. 1a, will be loaded with intact HBF and will be equipped with a modified lid to allow insertion of temperature probes inside the cask at various axial and radial locations. Design of the modified cask lid for instrumentation began in 2014, and a storage license will be obtained from the NRC. Cask loading of HBF is planned for 2017.

The specific guidance and recommendations of the ISG documents are discussed in Element 2, "Preventive Actions," in AMP M5. These ISG documents are considered as sources of preventative or mitigative actions that provide assurance that the structural and functional integrity of the canister internals will be maintained consistent with its design basis during the period of extended operation. For example, ISG-11, Rev. 3, focuses on the acceptance criteria to ensure integrity of the cladding as follows:

- For all fuel burnups (low and high), the maximum calculated fuel-cladding temperature should not exceed 400 °C (752 °F) for normal conditions of storage and short-term loading operations (e.g. drying, backfilling with inert gas and transfer of the cask to the storage pad). However, for low-burnup fuel, a higher short-term temperature limit may be used if an operator can show by calculation that the best-estimate cladding hoop stress is ≤90 MPa (13,053 psi) for the temperature limit proposed.
- During loading operations, repeated thermal cycling (repeated heatup/cool-down cycles) may occur but should be limited to fewer than 10 cycles, with cladding-temperature variations that are less than 65 °C (117 °F) each.
- For off-normal and accident conditions, the maximum cladding temperature should not exceed 570 °C (1058 °F).

Casks for HBF may have cladding walls that have become relatively thin from in-reactor formation of oxides or zirconium hydride. For design-basis accidents, where the structural integrity of the cladding is evaluated, one should specify the maximum cladding-oxide thickness and the expected thickness of the hydride layer (or rim), which may not be uniform. Cladding-stress calculations should use an effective cladding thickness that is reduced by those amounts and that has been justified by the use of oxide-thickness measurements and valid computer codes.

The guidance in ISG-11, Rev. 3, for storage of HBF for an initial period of 20 years, was based on short-term laboratory test data and analysis, which may not be applicable to the storage of HBF beyond 20 years. A major concern addressed in ISG-11 was the potential detrimental effect of hydride reorientation on cladding integrity. Because of the presence of radial hydrides, HBF cladding could exhibit a ductile-to-brittle transition temperature (DBTT) that could influence the retrievability of HBF assemblies and result in operational-safety problems for HBF that has cooled below the DBTT (i.e. ≈200 °C or 392 °F) [5, 6]. Figure 2 shows the ductility comparison for high-burnup M5® cladding, as-irradiated versus following simulated-vacuum drying—that is, cooling at 5 °C/h from 400 °C at peak hoop stresses of 90, 110, and 140 MPa, with the corresponding DBTT of <20, 70, and 80 °C, respectively.

Figure 3 compares the hydride morphology in the high-burnup M5® cladding, showing the microstructure and extent of radial hydrides across the cladding thickness, which has a strong correlation to the observed DBTTs in Fig. 2.

ISG-24, "The Use of Demonstration Program as Confirmation of Integrity for Continued Storage of High Burnup Fuel beyond 20 Years," provides guidance for the storage of HBF for periods greater than 20 years; it supplements the ageing-management guidance given in the NRC (NUREG)-1927 report [7]. ISG-24 specifies that operators may use the results of a completed or an ongoing demonstration, in conjunction with an actively updated AMP, as an acceptable means for confirming that the canister or cask contents satisfy the applicable regulations. Given that limited AMP action (e.g. condition monitoring) can be taken inside a sealed canister, the AMP must ensure that the TLAA associated with the ageing

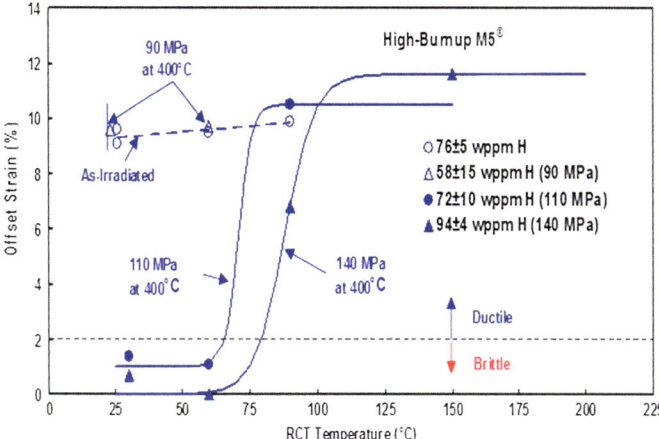

Fig. 2 Ductility comparison for high-burnup M5® cladding, as-irradiated versus simulated-vacuum drying (i.e. cooling at 5 °C/h from 40 °C at peak hoop stresses of 90, 110 and 140 MPa)

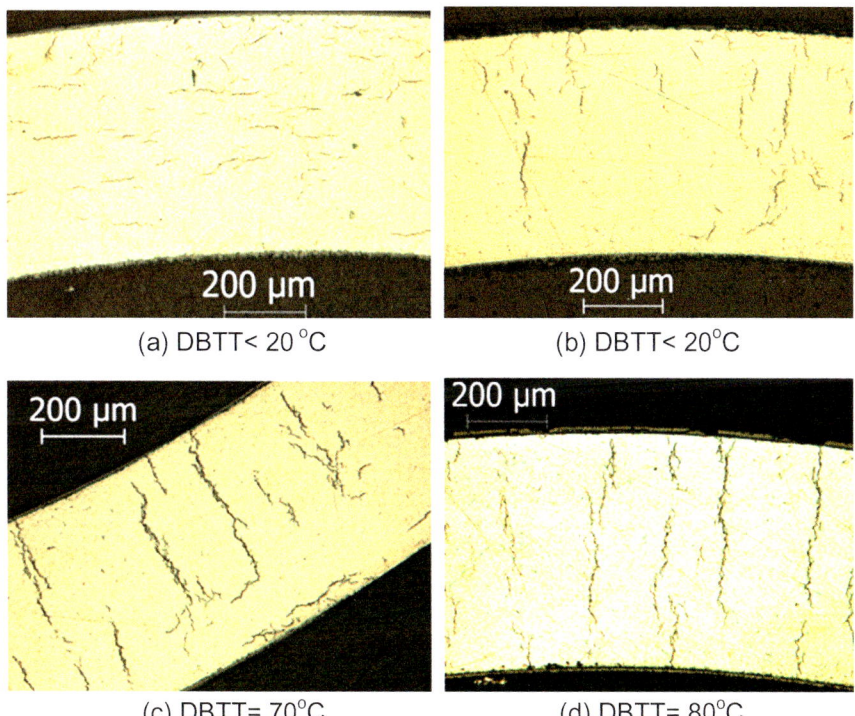

Fig. 3 Hydride morphology in high-burnup M5®: **a** as irradiated and **b**, **c**, and **d** following simulated vacuum drying (i.e. cooling at 5 °C/h from 400 °C at peak hoop stresses of 90, 110 and 140 MPa, respectively)

degradation phenomenon for HBF-cladding integrity is updated with new information as it becomes available. ISG-24 further specifies that the TLAA and AMP should be periodically re-evaluated and updated whenever new data from the demonstration or other short-term tests or modeling indicate potential degradation of the fuel or deviation from the assumptions of the TLAA or AMP. The updated TLAA and AMP should be reviewed by NRC for approval and will be subject to audit and inspection.

Because the boundary between existing knowledge and unknown factors or knowledge to emerge in the future always exists, *guidance*, in a way, is similar to *best practice,* which may be defined as a method empirically proven to yield excellent results and accomplish the stated objective. As expected, both regulatory guidance and industry best practice will evolve as knowledge increases and experience is gained over time.

2.2 Thermal Modeling

Thermal modeling of dry casks has been identified as a high-priority item in several gap analyses conducted by DOE and others. Best-estimate thermal modeling of temperature profiles of SSCs in casks/canisters is needed to understand vacuum drying, during which the HBF cladding may experience high temperatures and high internal pressure, resulting in high cladding-hoop stresses that would enhance hydride reorientation and cladding embrittlement during slow cooling [8]. Best-estimate thermal modeling is also needed to understand prolonged dry storage, given that the fission-product decay-heat load in the cask/canister decreases with time, because stress corrosion cracking of welds in marine environments is a relatively low-temperature threshold phenomenon associated with salt deliquescence [9]. Best-estimate thermal modeling is furthermore needed to benchmark the calculated temperatures against data measured, for example, in the High Burnup Dry Storage Cask Research and Development Project [4]. Finally, best-estimate thermal modeling is needed in particular to study canisters with welded closures to help determine the locations where temperature probes should be placed on the canister surface to detect potential breach(es) of the canister with a leakage of helium. Currently, there is no easy way of detecting the breach of a welded canister other than by condition monitoring of surface-temperature change, as was demonstrated, in principle, by Takeda et al. [10].

Three-dimensional (3D) simulation of thermal performance of a vertical dry-storage cask was conducted by Argonne researchers using ANSYS/FLUENT in 2014 [11]. Figure 4a shows a 3D model of one-fourth of a section in the vertical storage cask shown in Fig. 1b. The major components include a concrete overpack enclosed in carbon-steel shells, a multi-purpose canister (MPC) of welded stainless steel and a cask lid and a baseplate with radiation shielding and air inlet and exit vents that provide passive cooling of the MPC by natural convection. The inner and outer diameters of the overpack are 1.765 m and 3.366 m, respectively, and the

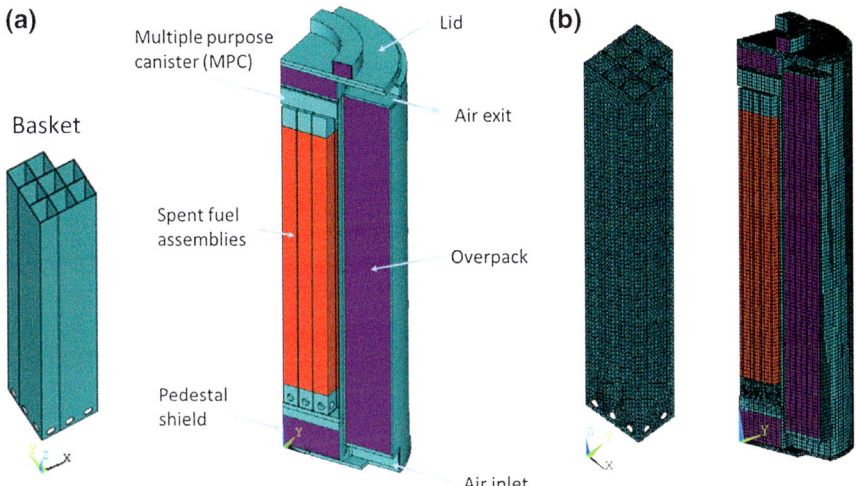

Fig. 4 **a** Cross section of one-fourth of a section of a 3D model of a fuel basket and cask showing major components and **b** the corresponding mesh representation. Spent-fuel assemblies are not shown for clarity

height of the overpack cavity is 4.864 m. Near the top and bottom of the overpack are four openings to allow air ventilation and canister cooling. The outer diameter of the MPC is 1.74 m, and the height is 4.82 m. The overall height of the cask is 6.083 m. At the bottom and on the top of each fuel assembly are a header and a footer/spacer combination, respectively. Figure 4b shows the mesh for the one-quarter section of the basket and cask (spent-fuel assemblies are not shown for clarity). The total number of cells is ~5.05 million. The grid was generated by using a meshing module in the ANSYS Workbench, which is a platform that facilitates meshing, solving (FLUENT) and data transfer in the 3D simulation. A high-fidelity mesh representation of the physical model can be obtained automatically by tuning the mesh control parameters in the workbench. In the case of fluid channels (e.g. radial and axial gaps through which natural convection occurs), they are represented by no less five cells across in the smallest channel.

The vertical cask contains a welded canister for 32 Pressurized Water Reactor (PWR) spent-fuel assemblies with a total maximum decay-heat load of 32.7 kW. To simplify thermal calculations, an effective thermal conductivity model for a used-fuel assembly was developed by using ANSYS/Mechanical. The effects of canister fill gas (helium or air), internal pressure (1–6 atm) and basket material (stainless steel or aluminium alloy) were studied to determine the peak cladding temperature (PCT) and the canister surface temperatures (CSTs). Table 2 shows the 3D simulation results of the PCTs and CSTs following a small canister leakage, during which the pressure of the He fill gas drops from 6 to 1 atm, and the He is eventually replaced by air at ambient pressure. The calculated PCT increases from 387 to 462 °C when the pressure in the He-filled canister drops from 6 to 1 atm,

Table 2 The 3D simulation of peak cladding temperature and canister surface temperature [11]

Fill Gas	Pressure (atm)	PCT (°C)/(z/L)	CST (°C)[a]
He	6	387/(0.85)	178
He	1	462/(0.54)	96
Air	1	500/(0.71)	126

[a]The locations of CSTs are at the center top of the MPC (shown encircled in Fig. 5a, b, and c)

whereas the axial location of the PCT decreases from z/L = 0.85 to 0.54, which is only slightly above mid-plane of the canister (i.e. z/L = 0.5, where L is the canister height). The PCT increases further from 462 to 500 °C when the He fill gas at one atm is replaced by ambient air, and the axial location of the PCT increases from z/L = 0.54 to 0.71. These changes are generally consistent with what would be expected from first-principle physics of natural convection, geometry and fluid properties, particularly gas densities and pressure.

Figure 5 shows the CST contours for the three cases in Table 2. The calculated CST at the center top of the canister (shown encircled in Fig. 5) first decreased from a peak value of 178–96 °C when the canister fill gas changed from six-atm He to one-atm He; it then increased to 126 °C when the remaining He was replaced by ambient air. Notice that when the canister leaks, heat transfer by natural convection inside the canister becomes less effective because of depressurization, and the gas layer above the spent-fuel assemblies hinders heat transfer to the top lid. As a result, the decay heat in the fuel assemblies accumulates toward the center and mid-plane of the canister, showing increased PCTs but with the peak value shifting downward, as shown in Table 2. The locations of the peak CSTs also change from the center top of the canister with six-atm He (Fig. 5a), to the side with one-atm He (Fig. 5b) and ambient air (Fig. 5c).

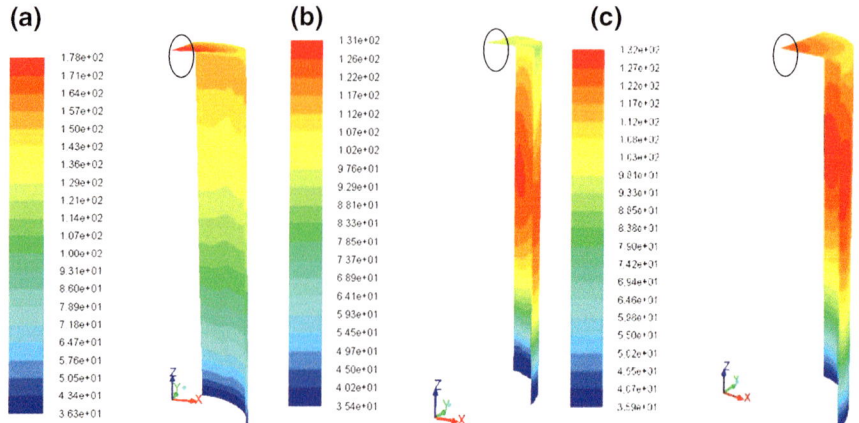

(a) **(b)** **(c)**

Fig. 5 The 3D simulation of contours of canister surface temperatures: **a** six-atm He, **b** one-atm He and **c** one-atm air

The results of the 3D simulation of thermal performance of a vertical dry-storage cask showed that natural convection inside the canister has significant effects on both PCT and CST, as shown in Table 2. Of particular interest to monitoring is the identification of canister locations where significant temperature change occurs after a canister breach and the fill gas changes from high-pressure helium to ambient air. The 3D simulation has provided insight into the temperature profiles of key components in a vertical cask for storage of HBF, particularly the change in CST following leakage of helium from a welded canister.

2.3   Monitoring and Inspection

Monitoring the interior of a welded canister containing spent-fuel assemblies for its functional and structural integrity is exceptionally challenging because of the intense levels of heat and radiation and the difficulties of transmitting the sensor signals out through the sealed stainless-steel canister wall. Yet, confirmation of canister integrity is crucial for the ageing management of the dry-cask storage systems during extended long-term storage and subsequent transportation of spent fuel. A canister breach can lead to serious consequences: release of radioactive contaminants; oxidation of fuel cladding, which could compromise fuel rod integrity and criticality safety; and generation of potentially explosive hydrogen gas.

Argonne researchers have developed an advanced surveillance technology called ARG-US (meaning "watchful guardian") Remote Area Modular Monitoring, or RAMM, for deployment in critical nuclear facilities, including DCSSs at the ISFSI sites across the country [12–14]. Coupled with thermal modeling of a dry cask, RAMM should be able to confirm canister integrity by continuously measuring the exterior surface temperatures of the canister—a much more manageable task than monitoring the interior conditions of the canister. The fundamental premise is that, when the canister breaches, its original pressurized helium cover gas would leak out, thus altering the established convective heat-transfer pattern within. As a result of the decreased upward convective flow, the canister top would become cooler. This temperature decrement is on the order of tens of degrees Celsius for a vertically-oriented dry-storage cask containing HBF, according to the simulation results shown in Table 2 and Fig. 5, as well as the surface-temperature measurements from the early experiments [10].

Figure 6 shows the layout of the major components in a weather-tight RAMM enclosure suitable for a wide range of indoor and outdoor applications. In developing ARG-US RAMM, a key requirement was modularity [12–14]. For hardware, modularity requires that sensors (e.g. thermocouples, gamma and neutron detectors and accelerometers) and communication components (e.g. wired Ethernet and cellular, satellite, and wireless-sensor networks [WSN]) be placed on daughter boards that can be readily interchanged and reconfigured. With this strategy, all sensors are to be treated comparably as long as the differences in device driver are

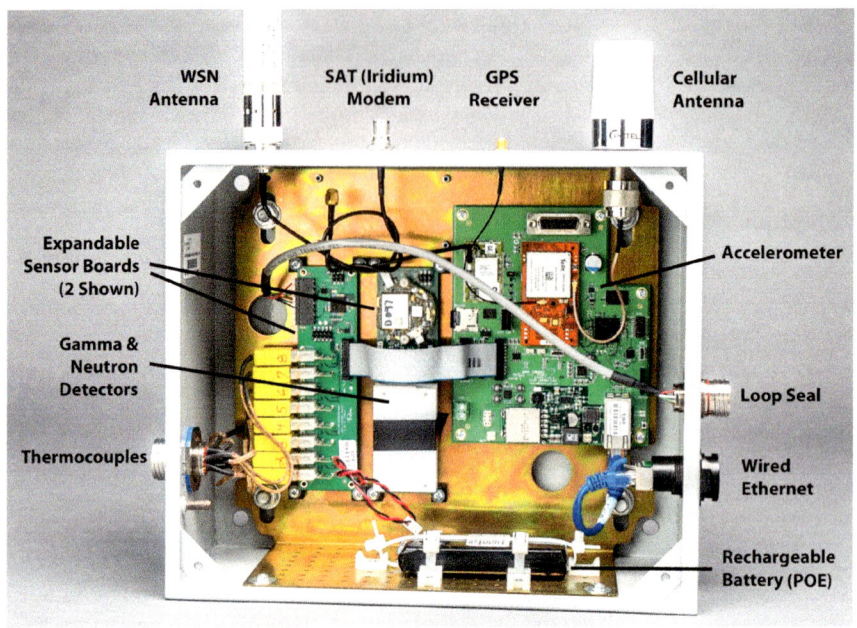

Fig. 6 Layout of major components in a RAMM unit

taken into account by the software; thus, any hardware combination can be handled. This modular approach reduces costs and increases application flexibility. Each sensor daughter board has its own sensor-interface processor and firmware programme to acquire and process the sensor data, including control of the alarms when readings are outside of the preset threshold values, both high and low for the majority of the sensors. By developing sensor interfaces that comply with the generic communications and interface formats of the RAMM motherboard, the number and type of sensors supported are practically unlimited. This methodology and the inherent flexibility associated with it allow continuous expansion of the system's utility and application base.

The WSN topology selected for the ARG-US RAMM system is that of a distributed "mesh," one that enables the units, or nodes, to communicate with any of their nearby neighbors in range [14], which is shown schematically in Fig. 7. The RAMM units would be distributed in such a way to allow for multiple routing paths between nodes, should a few units, or links, be damaged as a result of a disruptive event. This "self-configuring" or "self-healing" feature is a prerequisite for the successful deployment of a RAMM system. Near the edge of the network, some nodes would be given additional communication capabilities, such as cellular and Iridium satellite modems and global-positioning-system (GPS) receivers, to enable them to be the system "gateways" to the outside world—for example, to an off-site control center.

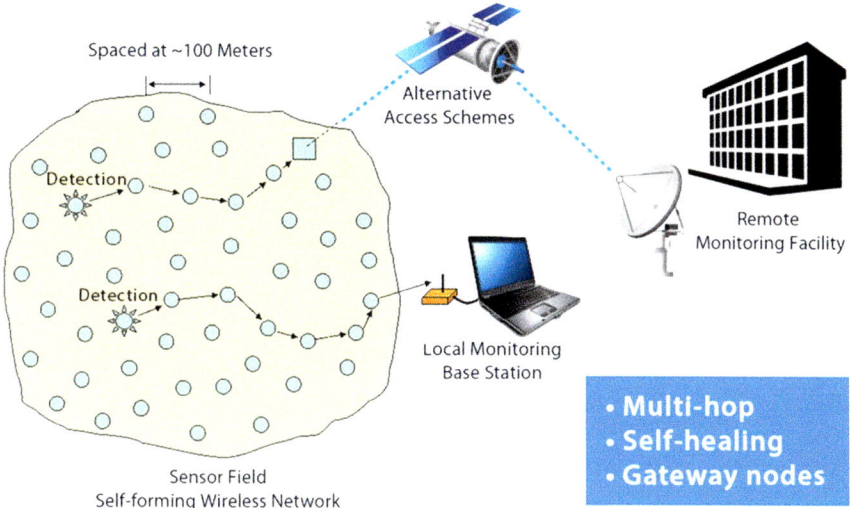

Fig. 7 Schematic illustrating a RAMM wireless-sensor network

In the accompanying RAMM software developed by the Argonne researchers, the concept of modularity is also incorporated [14]. A controller is used to poll the sensor values, process sensor alarms and forward the data to the communication manager. A translator and data logger then format and locally store the sensor events and data in a uniform format. From the direction of commands and requests, the communication manager and an additional translator allow for situational modifications of the data before transmission and after reception, as well as incorporations of encryption and authentication algorithms. A block diagram of ARG-US RAMM and the software structure, as well as a conceptual layout of the RAMM components, can be found in reference [14].

Development of a basic RAMM unit is largely complete; production of a limited number of prototype units has begun, and they should be available for field-testing in the summer of 2015. Figure 8 shows a conceptual view of field-testing and evaluation of multiple (e.g. 5–10) RAMM units, each fitted to a dry-storage cask on an ISFSI site. Multiple RAMM units are recommended for field-testing because of the need to form a wireless-sensor network for performance evaluation and, even more importantly, for trending—allowing comparison of canister surface temperatures measured for each installed dry- storage cask against its known heat load under nearly identical environmental exposure conditions. Whereas leakage of an individual canister may be difficult to decipher—particularly if the leak rate is low—trending may be possible in making any perturbations more visible with a populated database of five or more canisters. Daily and seasonal variations in ambient temperature and blockage of vents on the dry casks are factors that would affect canister surface temperatures; the changes in patterns should be readily identifiable from the histograms of the recorded data.

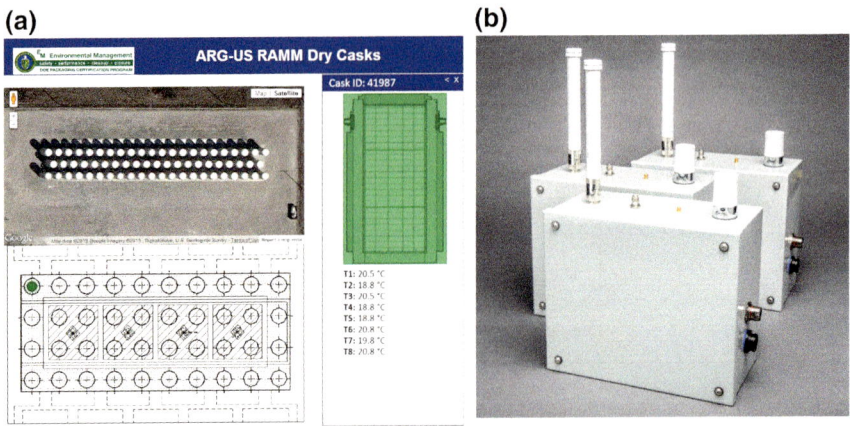

Fig. 8 Conceptual view of field-testing and evaluation of **a** multiple RAMM units, **b** each fitted to a dry-storage cask on an ISFSI site

A weather-protected RAMM unit, such as those shown in Fig. 8b, would be installed for each vertical dry-storage cask with thermocouples (TCs) routed through the air-exit vent (shown in Fig. 1b) and placed near the top center and side of the canister, where large changes in CSTs are predicted to occur after a canister breach. Each RAMM unit would have its own power supply from the grid and battery backup (shown in Fig. 6), a suite of sensors (e.g. temperature, humidity, shock, radiation [gamma and neutron] and seal) and data and alarm communication capabilities. Multiple RAMM units are configured into a wireless-sensor network that is self-forming, or self-healing, even after certain units are damaged and no longer fit for service. It is also envisioned that once the RAMM units are properly placed on dry casks, they are programmed to stay dormant, except during the brief periods of scheduled measurements. Being outside of the vertical storage cask and away from the spent-fuel assemblies, the sensors and electronic boards in the RAMM units are not directly exposed to the sources of radiation and heat and, therefore, should operate reliably during extended service. As Fig. 8a shows, a secure webpage of ARG-US RAMM for dry-storage casks on an ISFSI site has already been developed, along with secured backend data servers.

3   Other Ageing Management Programs and Discussion

Chapters 3 and 4 of the ageing management report [1] contain five TLAAs and seven AMPs, respectively, for managing ageing effects on the structure, system and components in the DCSSs and ISFSIs during extended long-term storage and subsequent transport of used fuel. Summary descriptions and elaboration of these AMPs and TLAAs can be found in references [15–17]. Among the seven AMPs, two—the *Welded Canister Seal and Leakage Monitoring Program* (AMP IV.M3) and the *Canister/Cask Internals Structural and Functional Integrity Monitoring Program* (AMP IV.M5)—are related to welded canisters and depend on monitoring and inspection to detect ageing effects on welded canisters and canister internals. Both AMPs recommend following the guidance in NRC ISG documents as preventive actions to help ensure proper confinement evaluation, vacuum drying, material evaluation, design and testing of lid welds and pressure and helium leakage testing of the confinement boundary of spent-fuel dry-storage systems. Both AMPs consist of three main parts: (1) assessment of records of canister fabrication, testing and fuel loading operations pertaining to the relevant ISG documents; (2) thermal modeling; and (3) monitoring and inspection. Early detection of ageing effects on structures, systems and components before the loss of their intended safety functions occurs is covered in Element 4 in AMPs IV.M3 and IV.M5; monitoring and trending is covered in Element 5 in AMPs IV.M3 and IV.M5. Element 6 in AMPs IV.M3 and IV.M5 addresses acceptance criteria, whereas Elements 7, 8 and 9 in AMPs IV.M3 and IV.M5 address, respectively, corrective actions, confirmation process and administrative controls. Element 10 in AMPs IV.M3 and IV.M5 addresses the operating experience of DCSS designs at existing ISFSI sites as knowledge continues to accumulate going forward.

As shown elsewhere [11] and in this chapter, best-estimate thermal modeling of dry-storage casks containing spent-fuel assemblies enables assessment of PCTs and CSTs against certain temperature thresholds, such as: (1) 400 °C for potential hydride reorientation in high-burnup PWR cladding alloys [8]; or (2) 85 °C for the temperature below which chloride-induced stress corrosion cracking (CISCC) can potentially degrade a stainless-steel canister weld in a marine environment as a result of salt deliquescence and the residual stress in the weld [9]. The temperature profiles of spent-fuel assemblies and canisters obtained in the 3D simulation depend on many factors; they are not static and will change during prolonged dry storage as the fission-product decay-heat load in the canister decreases with time. However, once the model for the dry-storage cask is constructed, conducting 3D simulations for the changes in the cask heat load and/or ambient environment is straightforward.

Note that managing ageing effects on DCSSs for extended long-term storage and subsequent transportation of used fuel "begins" when the used-fuel assemblies are loaded into a canister (or cask) under water in the spent-fuel pool. The canister (or cask) containing the used-fuel assemblies is then drained, vacuum dried and back-filled with helium to provide an inerting environment for long-term storage. The bolted cask, as well as the welded canister (after being placed inside a transfer cask), are moved to an outdoor concrete pad of an ISFSI, where it will stay for 20 or 40 years of the initial license term (and up to another 40 years for a renewal license term), according to 10 CFR 72.42. More than 2,000 dry casks have begun long-term storage under the initial license terms; some of them have been in storage for more than 20 years and are already in the renewed license term of up to an additional 40 years. Transferring from pool to pad, or from wet to dry storage, represents an abrupt change of environment for the used-fuel assemblies, and the effects of this change are most pronounced during vacuum drying, especially for

HBF, because of the likelihood of cladding radial- hydride formation and embrittlement. The likelihood of this phenomenon will diminish only after the cladding temperature has dropped below 200 °C; because of the decrease of fission-product decay heat during prolonged cooling, which may occur 20–25 years after the high-burnup, used-fuel assemblies are placed in dry storage. Preventing and/or minimizing cladding embrittlement by radial hydrides during drying, transfer and early stages of storage will maintain the configuration of the used fuel in the dry canister (or cask) and ensure retrievability of the used fuel and its transportability after extended long-term storage.

Management of ageing effects on DCSSs for "extended" long-term storage of used fuel is no different from that required during the "initial" license term. If ageing effects on the SSCs important to safety in the DCSS/ISFSI are not adequately managed for the initial license term of storage, an application for a renewal of the license for extended long-term storage is unlikely to be granted by the regulatory authority. Therefore, the same principles and guidance developed by the NRC in NUREG-1927 should be applicable to extended long-term storage, as the period of operation, or term, reaches 20, 40, 60, or >80 years. The term in the initial or renewal license is important and indicates a finite period of operation and, although not mentioned specifically in the current regulations, this does not rule out license renewal for multiple terms, as long as ageing effects are adequately managed.

The main goal of the ageing-management report [1] is to provide a framework to help establish the technical basis for extended long-term storage and subsequent transportation of used fuel. Potential future activities include the development of additional AMPs and TLAAs, as necessary, and further evaluation of the adequacy of the generic AMPs and TLAAs that may need augmentation. Industry and site-specific operating experience that has been captured/logged from the various DCSSs/ISFSIs located across the USA should be reviewed periodically to: (1) ascertain potential ageing effects on the SSCs in the DCSSs, thereby enabling a compilation of existing aging management activities; and (2) assess the adequacy of these activities for extended long-term storage and transport of used fuel. The US industry has incorporated the ageing management report [1] into Nuclear Energy Institute (NEI) 14-03: "Industry Guidance for Operations-Based Aging Management for Dry Cask Storage" [18] to augment Title 10 of the CFR Part 72 license and the Certificate of Compliance (CoC) renewal review guidance in NUREG-1927 [7], "Standard Review Plan for Renewal of Used Fuel Dry Cask Storage System License and Certificate of Compliance," Sect. 3.0, "Aging Management Review." NEI 14-03, Revision 0 was submitted for NRC endorsement in September 2014; the NRC is updating NUREG-1927, and the draft was supposed to be available for public comment in the summer of 2015. (Note: NRC has updated NUREG-1927, Revision 1 in June 2016 and changed its title to "Standard Review Plan for Renewal of Specific Licenses and Certificates of Compliance for Dry Storage of Spent Nuclear Fuel.")

Note that NEI 14-03 introduced a concept called "tollgates" into the Part 72 license and CoC renewal process. "Tollgate" is a new term created by the nuclear industry to address the fact that the applicability of potential dry-cask-storage ageing mechanisms may not be able to be verified at the time that the license and CoC renewal applications are submitted. This information would enhance the current understanding of the future state of dry spent fuel and the canisters that contain it. Tollgates are part of a learning, operations-based ageing management programme implemented by licensees via requirements in the renewed license or CoC and the associated final safety analysis report. These requirements obligate the licensees to perform periodic assessments of the aggregate state of knowledge of ageing-related operational experience, research, monitoring and inspections to ascertain the ability of in-scope DCSS-design SSCs to continue performing their intended safety functions throughout the requested period of extended operation.

4 Conclusion

Managing ageing effects on DCSSs for extended long-term storage and transportation of used fuel requires knowledge and understanding of the various ageing degradation mechanisms for the materials of the SSCs and their environmental exposure conditions during the extended period of operation. The operating experience involving the AMPs, including the past corrective actions resulting in programme enhancements or additional programmes, provides objective evidence to support a determination that the effects of ageing will be adequately managed so that the intended functions of the SSCs will be maintained during the period of extended operation.

Managing ageing effects on DCSSs for extended long-term storage and transportation of used fuel depends on AMPs for the prevention, mitigation and early detection of ageing effects on the SSCs through condition and/or performance monitoring. Detection of ageing effects should occur before there is a loss of any structure's or component's intended function that is important to safety. Among the important aspects of detection are the method or technique employed (i.e. visual, volumetric or surface inspection), frequency, sample size, data collection and timing of new/one-time inspections to ensure timely detection of ageing effects. The challenges in the detection of ageing effects include: (1) detection of ageing in regions that are inaccessible for inspection and monitoring; and (2) determination of the frequency of inspection and monitoring (i.e. periodic versus continuous). The DOE, the NRC and industry personnel worldwide are actively engaged in various R&D and demonstration activities and meet regularly to exchange information through EPRI's Extended Storage Collaboration Program and other venues, such as International Atomic Energy Agency meetings and international conferences.

Acknowledgements This work is supported by the U.S. Department of Energy's Used Fuel Disposition Research and Development, the Office of Nuclear Energy (NE) under Contract DE-AC02-06CH11357. The author wishes to acknowledge the contributions from his colleagues at Argonne National Laboratory: D.R. Diercks, O.K. Chopra, M.C. Billone, HC Tsai, J. Li, Z.G. Han, D. Ma, and R.R. Fabian.

References

1. Chopra OK (2014) Managing aging effects on dry cask storage systems for extended long-term storage and transportation of used fuel, Rev. 2, FCRD-UFD-2014-000476, 30 Sept 2014
2. NUREG-1801 (2010) Generic aging lessons learned (GALL) report, Rev. 2. U.S. Nuclear Regulatory Commission, Washington, D.C., Dec 2010
3. Einziger RE (2013) An aging management for spent fuel dry storage and transportation. Radwaste Solut July–August 2013
4. EPRI (2014) High burnup dry storage cask research and development project, final test plan. Electric Power Research Institute, Contract No.: DE-NE-0000593, 27 Feb 2014
5. Billone MC, Burtseva TA, Einziger RE (2013) Ductile-to-brittle transition temperature for high-burnup cladding alloys exposed to simulated drying-storage conditions. J Nucl Mater 433:431–448
6. Billone MC, Burtseva TA, Liu YY (2013) Baseline properties and DBTT of high-burnup PWR cladding alloys. In: Proceedings of the 17th international symposium on packaging and transportation of radioactive materials, PATRAM 2013. San Francisco, CA, 18–23 Aug 2013
7. NUREG-1927, Rev. 0 (2011) Standard review plan for renewal of spent fuel dry cask storage system licenses and certificates of compliance. U.S. Nuclear Regulatory Commission, Washington, D.C., March 2011
8. Billone MC, Burtseva TA, Liu YY (2013) Effects of drying and storage of high-burnup cladding ductility. In: Proceedings of international high-level waste management conference. Albuquerque, NM, 28 April–2 May 2013
9. Oberson G et al (2013) U.S. NRC-sponsored research on stress corrosion cracking susceptibility of dry storage canister materials in marine environments. Proceedings of 39th waste management conference, WM2013. Phoenix, AZ, 24–28 Feb 2013
10. Takeda H et al (2008) Development of the detecting method of helium gas leak from Canister. Nucl Eng Design 238(5):1220–1226
11. Li J, Liu YY (2014) Thermal performance of a vertical dry cask for storage of high burnup used fuel. In: Proceedings of 55th annual meeting of institute of nuclear materials management. Atlanta, GA, 20–24 July 2014
12. Tsai HC, Liu YY, Nutt M, Shuler JM (2011) Advanced surveillance technologies for used fuel long-term storage and transportation. In: Proceedings 14th international conference on environmental remediation and radioactive waste management. Reims, France, 25–29 Sept 2011
13. Tsai HC, Liu YY, Shuler JM (2013) Monitoring critical facilities by using advanced RF devices. In: Proceedings of the 15th international conference on environmental remediation and radioactive waste management. Brussels, Belgium, 8–12 Sept 2013
14. Tsai HC, Craig B, Lee H, Mittal K, Liu YY, Shuler JM (2014) ARG-US remote area modular monitoring for dry casks and critical facilities. In: Proceedings of INMM 55th annual meeting. Atlanta, GA, 20–24 July 2014
15. Diercks DR, Chopra OK, Ma D, Liu YY (2014) Holistic aging management for extended storage and transportation of used nuclear fuel. In: Proceedings of INMM 55th annual meeting. Atlanta, GA, 20–24 July 2014

16. Chopra OK, Diercks DR, Ma D, Han ZG, Liu YY (2014) Role of time-limited aging analysis in managing aging effects on used fuel dry storage systems. Proceedings of INMM 55th annual meeting. Atlanta, GA, 20–24 July 2014
17. Diercks DR, Ma D, Chopra OK, Liu YY (2014) Degradation of concrete structures in used nuclear fuel dry cask storage systems. In: Proceedings of INMM 55th annual meeting. Atlanta, GA, 20–24 July 2014
18. Nuclear Energy Institute (2014) Industry guidance for operations-based aging management, NEI 14-03. 23 Sept 2014

Part V
The Aesthetics of Ageing

Forever Young or Ageing Naturally?

Rob P. J. van Hees and Silvia Naldini

Abstract Age and ageing can be felt as negative occurrences. For monuments, however, old age is traditionally considered to be a positive quality. Without a certain age, the designation "monument" hardly applies. Ageing can be seen as the work of time, which has always been valued: ageing was sometimes even artificially induced in the past. In this paper, we will discuss the meaning of ageing in monumental buildings. The fact that, in the case of interventions in monuments, a perpetual service life is strived for, but restoration ethics clearly put limitations on what can be done can lead to dilemmas and can make it difficult to make decisions. Cases will be discussed to derive some criteria to base interventions upon, seeking a balance between merely preventive conservation and rejuvenating practices.

Keywords Monuments · Ageing · Preventive conservation

1 Introduction

Ageing is often felt as a scary and negative process. For monuments, however, old age is traditionally a positive quality: without being of a certain age, a building will hardly be listed as a monument. Ageing and patina are concepts whereupon many discussions in the field of architectural conservation focus. One of the first questions to be answered concerns the meaning of ageing and patina.

Ageing of monuments is a natural process, which may result in the formation of patina. A definition of ageing is 'the process of growing old or developing the appearance and characteristics of old age' [1]. Patina can be also considered the result of a chemical transformation of the surface (only) of the material. As a

R. P. J. van Hees (✉)
TNO Technical Sciences, Delft, The Netherlands
e-mail: r.p.j.vanhees@tudelft.nl

R. P. J. van Hees · S. Naldini
Faculty of Architecture and the Built Environment, Delft University of Technology, Delft, The Netherlands

© Springer International Publishing AG 2018   141
K. van Breugel et al. (eds.), *The Ageing of Materials and Structures*,
https://doi.org/10.1007/978-3-319-70194-3_11

general term, patina refers to the change in an object's surface resulting from natural aging [2]. The removal of the patina needs careful consideration, as it can lead to the obliteration of the traces of time. Ageing could be accelerated by exogenous decay processes—like salt or frost-damaging mechanisms—affecting the cohesion of the materials of a monument. A decay process can also develop inside the ancient materials, enhancing their ageing dramatically, without any environmental aggression. This is the case of creep, affecting the structural strength of the building. Ageing due to damaging processes is obviously unwanted and should be hindered.

One could further wonder whether an aged look could even be desired and patina could be artificially created. In regard to paintings, the discussion about the wanted patina is centred on the procedures assumed to be used by old masters to give their colours homogeneity, starting with Apelle's 'atramentum' reported by Plinius [3]. Stone statues and façades could be treated, as well, which should be considered when cleaning is being considered. In the case of painted masonry, the colour will also be altered as a result of ageing, which means that various materials will show various different signs of ageing.

Some cases of ageing will be commented upon, aiming at exploring the meaning of ageing and extracting some values to refer to when confronted with the dilemma: allow ageing or intervene? and also when deciding on the extent of the intervention. We will start with a ruin, an extreme case of ageing, to further tackle the problems of the extent of the thoroughness, and the envisaged result, of cleaning and restoration actions. How to approach problems of restoration is a complex matter. In his Teoria del Restauro, first published in 1963, Cesare Brandi, director of ICR (Central Institute of Restoration) founded in 1939 in Rome, faced most of these problems, which have been debated up to the present day. It is therefore interesting to recall some of his principles in our discussion of the cases, to see whether they can be referred to in the light of the present quest for a new balance between *young* and *old*.

2 The Castle Ruin of Asten (the Netherlands)

A ruin is any object giving witness to human history, even though not having been preserved in its original form, which may be even hard to recognise as one. In the case of the Castle of Asten (the Netherlands) [4], only a ruin remains, which however shows interventions made at various times in the past. The castle dates back to 1430 ca. and went through transformations over the course of time, the latest concerning the fantasy interventions of Baron Clemens van Hövell tot Westerflier in the 20th century. The current attitude is to keep each addition to the original body, even the most recent one, dating to the first decennia of the 20th century, because it is perfectly recognisable as such and provides added historic value to the object. The initial form of architecture at the moment of its creation should no longer re-created, but all its components should be maintained as expressions of different techniques and craftsmanship belonging to different historic phases. From the hand-moulded brick to the concrete, from the lime mortar to mortars containing cement, all materials and techniques are individual expressions of historic periods, and all have withstood time in various different ways. The orange bricks, originally meant for interior spaces and presently exposed to environmental agents, suffer from the exposure, in contrast to the darker bricks. The masonry of the 15th-century construction still shows a rather good state of preservation. Some maintenance needs to be done regarding re-pointing and solving the problems related to water penetration. Rather than the preservation of the *original* form and aesthetic value of the castle, the preservation of its historical value can be achieved by preventing restoration, that is to say, trying to maintain the current status quo [5]. Also, the environment, i.e. the countryside the castle lies in, if kept untouched, will provide the monument with the natural space that is necessary for its appreciation. A new aesthetic value will derive from this approach.

The foundation in charge of the building has decided to leave the ruin as it is, without integrations and limiting intervention to the essential, like substituting the pointing where necessary, which can be described as the conservation of the materials in their current state. Even the vegetation is meant to be maintained, that is to say, to be left growing onto the walls, only preventing it from becoming too heavy and inducing local collapse. This is an interesting approach, which is not aimed at guaranteeing the ruins *eternal life* and can be even called romantic. We would maybe act differently if the ruin was an ancient Roman excavation. The point

Fig. 1 In the ruin of the castle, the action of time can be seen, as well as the traces of historical events, including the grenade which hit the construction in 1944, starting the final phase of the castle in which the owner no longer resides (Photo Silvia Naldini)

would then be how to preserve it. Preserving the history involved in the ruin could also suggest a solution like the creation of a structure enclosing the ruin, as was done in the case of the Museum of the Roman Baths in Heerlen (the Netherlands) (Fig. 1).

The ruin would be separated from its historic environment and inserted into a new, artificial one. This is the alternative to Brandi's 'preventive restoration,' as pointed out by Urbani [6], who concludes noting that Brandi's theory is rather than a conclusive work, a first step to the future of restoration. The case of the ruin is an extreme one, and we are often confronted with problems concerning the preservation of the historical and aesthetic value of a monument still having its unity, even though its original form could have been altered by interventions. Preserving the building with the traces of the passage of time is a matter of careful analysis in each case. Only thorough study of the monument and the message attached to its forms could guide a good intervention.

3   San Marco's Campanile in Venice (Italy)

As mentioned above, certain damage mechanisms can enhance ageing dramatically. The following case of creep is pointed to as an example of ageing, finally leading to loss of the original form of the object.

The campanile (bell tower) of San Marco in Venice was constructed between 1156 and 1173. If we compare the sudden collapse of the San Marco bell tower on 14th July 1902, with more recent disasters like the Pavia tower, for which creep has been assessed as the major cause of failure, we may assume that a similar phenomenon may have occurred in Venice with San Marco's bell tower. Creep is a form of ageing. Within a few years after the collapse, it was decided to reconstruct the tower, and a new one was erected with the same proportions and materials as the one that collapsed ('as it was' and 'where it was') in 1912. The order of magnitude of creep of the historic masonry can be considerable and may therefore be a serious problem [7]. Creep in compression, due to dead loads, generally leads to (deep or trans-sectional) vertical cracks. This type of damage (passing through cracks) is typical of slender structural elements like stone or brickwork columns and piers and of heavy, but tall structures, like towers (and heavy structures to be found, for example, in ancient churches). It may develop within a relatively short or very long time, depending on the brittleness of the material and is due to the creep behaviour of the material when stressed beyond the elastic limit. Cracks can propagate very slowly for decades or even centuries, but, in the end, if the phenomenon is not stopped, the element or structure can suddenly collapse (Figs. 2 and 3).

San Marco's tower can be considered as a fundamental part of a historic site. The loss of this tower meant therefore that something similar had to come in its place. Also, Brandi points at this reconstruction as a way of re-establishing a unity, however, arguing that what was originally there and was lost cannot be revived: a copy is both a historical and an aesthetic falsification, unless made for didactical

Fig. 2 San Marco's bell tower, assumed behaviour during collapse (*Source* vecchie immagini di venezia – archivio filippi)

purposes [5]. In the case of Venice the essential missing element was a vertical body. The current tower is, except for the materials, maybe too much a copy of the lost one, which points at a *re-creation*. One could, however, argue that the time elapsed between collapse and re-erection was so short and the shock to the involved community so great that this reconstruction might have been justified.

4 The Colour of the Façade: The Restoration of St. Peter's in Rome (Italy) and the Royal Palace in Amsterdam (the Netherlands)

The problem of where to stop in a restoration process was faced in the following cases of cleaning and restoration. The relevance of each of the buildings and its representative function explain why the broad discussion originated.

Within the framework of the activities around the Jubilee Year 2000, having as its object Rome and its buildings, the façade of St. Peter's was restored. The surfaces were not only cleaned but also painted in the colours originally chosen by Maderno, the architect who extended the basilica's design nave and façade. Differently from many other cases, the colours were the means to almost theatrically create a depth in the façade, letting the white columns emerge, whereas the surface

Fig. 3 San Marco, Venice,
view of the church with the
reconstructed bell
tower (Photo Rob van Hees)

behind, in the nuances of tobacco–ochre, recedes. Maderno solved thus the problem
of the deviation from the design of Michelangelo's church and façade
re-establishing the form of the temple in an almost Baroque way. In this case, the
restitution was not merely meant to go back to the original colour given to the
materials, in contrast with the effects of the ageing process, but aimed at restoring
the unity and meaning of the original creation. Being such a prominent and symbolic building, the restoration of St. Peter's divided scientists and the general public
into two factions of supporters and opponents of the chosen approach. The party
against the restoration would use the argumentation that going back in time is not
only historically but also aesthetically incorrect, even though it could be demonstrated that traces of brown colour had emerged after removing the dirt layer,
brushes had been bought at the time of Maderno, and a painter had been paid for
colouring the wooden model of the façade to show how the final effect would look.
Too many façades in Rome, it was said, have been restored over the last decennia,
going back to the original light, pastel colours, instead of keeping the ochre-red
shade, which has become the dominant colour of Rome in the course of time.
Moreover, the opponents found the context important and that the colour of the
façade should match the colours of the buildings of the area. A final problem was
psychological or maybe emotional and consisted of the difficulty people have

Fig. 4 Façade of St. Peter's before restoration, 1985 (*Source* http://www.kofc.org)

adjusting to change. All mentioned elements needed to be taken into account and weighed against the elements in favour to reach a well-balanced and well-argued decision (Figs. 4 and 5).

The Dutch Government Building Agency aims at achieving well-conceived restorations based on respect for the traces of time [8]. Facing the problem of the removal of the patina formed on natural stone, an interesting approach was chosen for the restoration of the façade, including marble reliefs, of the main pediment of the Royal Palace in Amsterdam, originally built as the town hall of the city in 1648. Like St. Peter's in Rome, the palace is a prominent building for Amsterdam and the Netherlands, as testified by the genuine amazement of contemporaries like the poet Vondel, who named it 'the eighth world wonder'. The masterpiece of the great Dutch architect, Jacob van Campen, the building showed his interpretation of classical architecture and was the symbol of the power of the city of Amsterdam during the Dutch Golden Age. It is presently the most important Royal Palace of the Netherlands, which explains why its conservation has become the subject of a widespread controversy. The final result can be described as a well-argued restoration, reflecting awareness of the importance of perception [9]. One important characteristic of the palace is that its façade is made of sandstone, in a country where, due to the scarce availability of natural stone, the traditional building material is brick, and only the most relevant buildings are *clad* in natural stone. The sculptures on the pediments of both the front and rear façade are carved in Carrara marble. Both natural stone types had undergone ageing, resulting in a patina on the stone, as well as local staining. As far as the relief of the front façade is concerned, this was considered disturbing and hindered a thorough appreciation of the features

Fig. 5 Front façade St. Peters after intervention (Photo Rob van Hees, April 2000)

of the works of art. The condition of the façade was perceived as shabby and neglected, within the context of the Dam square and the surrounding buildings. The main problem was how to clean the stone. Going back to the original colour of the stone, removing *all* traces of time, would mean considering time as reversible, forgetting the historical context [5]. Besides, it was not clear how Van Campen had intended to make the stone blocks look more homogenous, whether he wanted them to be painted, following a common practice, or oiled, as it was proven to have been the case, in 1689.

The blocks had originally been tooled, and this was a relevant aspect to be maintained, because all parts originally formed a unity. Also, the cleaning technique was a matter of discussion, considering that the materials used were diverse, even the sandstone was of two types, Obernkirchen and Bentheim, which have different (ageing) characteristics, and the technical state of conservation of the stone could be critical [9]. The most suitable technique was chosen on the basis of a study and try-outs (Fig. 6).

It was decided to limit the intervention to a minimum, and not to remove all traces of patina, as not only the colour but also the texture would have been otherwise affected, and this would look artificial and awkwardly 'young'. Therefore, it was considered necessary to add a little colouring, or artificial ageing, for recreating the aesthetic unity of the building. The guiding principle was that any intervention needed to be reversible, and its effect should not vary in relation to the stone it was applied upon [8]. In case of poor quality of some of the stone blocks,

Fig. 6 Royal Palace Amsterdam, front façade and surrounding buildings (Photo Rob van Hees, before restoration in 2006)

repair work with a repair mortar was carried out; re-pointing was done when necessary. Returning to the issue of colour, the contrast was sought between sandstone walls and the marble decoration of the main pediment, which had been regularly painted with a lime wash until the 18th century. The cleaning and local consolidation were preceded and followed by a thorough documentation campaign aiming at documenting the state of the materials before and after the intervention (Fig. 7).

In conclusion, the idea of *undisturbing* was introduced, concerning the extent of the cleaning process. Leaving the patina was unacceptable, however too light a colour of the stone would have been disturbing. The removal of patina was therefore not integral, leaving some traces of its respectable age, and the stone parts that had been too light (for example, due to previous replacement) were made a bit darker, making the overall aspect of the façade more readable, but still ancient in aspect and homogenous in colour. Most outspoken was the cleaning of the pediment in Carrara marble, for which readability was the main aim and which is now again looking 'as new'. Still, the challenge was not to go back to the origin of its construction, but to make it readable.

Although each case is different, we can say that a general well-considered approach should not be aimed at going back to the moment of the creation of the

Fig. 7 Royal Palace Amsterdam, front façade after restoration (Photo Rob van Hees, Feb. 2014)

object, *cancelling* the time which separates us from then, but it should also not neglect the uniqueness and significance of the monument. There will be cases, like the façade of the St. Peter's, where the final result could be perceived as rejuvenating, even though, the restitution of colour means restitution of depth, of a vision of the architecture as true to the time, as it best as it could be. Already in 1978, Brandi agreed upon the necessity to clean the façade of St. Petronius in Bologna [10] not only because it would be justified by the then existing damage to the materials, but especially because the deposition layer on the surface hindered the appreciation of the two main colours of the stone used in the façade, which had become monochrome.

5 Discussion and Conclusions

Ageing and service life are closely connected issues with all buildings and building materials. Whenever a building does not fulfil its function anymore, or no longer possesses aesthetic qualities, the end of service life approaches and demolition is approaching. A monument is a different case because we want to preserve it, and we tend to strive for a kind of perpetual service life [11]. The question is then: may the monumental building show its age? And the answer to that is not 'yes' or 'no'—it is not black or white. If all buildings and all building materials age, then also so do monuments. The difference is that in ordinary buildings it is easy to decide on either maintenance and face-lifting or, in the most extreme case, on demolition, depending on the tastes and the wish of the owner.

In case of (listed) monuments, instead, there are many more people that may decide on a suitable approach, and there is no rule, at least no fixed rule; only demolition is normally impossible or at least only allowed when safety is severely compromised. The opposite, that is to say 'reconstruction' or rather a complete 'recreation', however, is also possible, even though severely debated, as mentioned in the case of San Marco's bell tower. 'Don't touch any of my wrinkles it took me *ages* to get them' actress Anna Magnani told her makeup artist—growing old gracefully [12]. This has been the attitude shown by Dutch and other EU heritage authorities towards the ageing of monuments for a long time. Nowadays, things tend to change, as awareness has grown that a monument is at least not fully comparable with the human body. The most important restoration of the past ten years in the Netherlands, that of the Royal Palace in Amsterdam, introduced concepts like *undisturbing*, for making too-light parts a bit darker and too-dark parts a bit lighter, thus leaving some traces of the respectable age, while making the overall aspect of the façade more equilibrated and better understandable and consequently enjoyable. More prominent was the cleaning of the pediment in Carrara marble, for which *readability* was the priority and which is now looking 'as new', showing that, even within the same monument, various parts could require different approaches. The choice needs to be supported by qualified and well-balanced judgement, enabling to act respectfully towards old age, because, as Brandi pointed out, no universally applicable solution exist.

References

1. http://www.thefreedictionary.com/ageing. Accessed Dec 2014
2. http://www.cartage.org.lb/en/themes/arts/scultpureplastic/UnderstandingSculpture/Patina/WhatisPatina/WhatisPatina.htm. Accessed Dec 2014
3. Plinius. Nat Hist XXXV, 41–43
4. http://kasteelasten.nl/historie/. Accessed 21 Jan 2014
5. Brandi C (1977, 1963) Teoria del restauro. Einaudi, Torino, pp 30–37
6. Urbani G (2000) Intorno al restauro. In: Zanardi B (ed). Skira, Genova-Milano, pp 69–75
7. van Hees RPJ, Binda L, Papayanni I, Toumbakari E (2007) Damage assessment as a step towards compatible repair mortars. In: Groot C, Ashall G, Hughes J (eds) Characterisation of old mortars with respect to their repair—RILEM report 28, pp 105–150. ISBN: 978-2-912143-56-3
8. van Bommel B (2008) De gevels van het Koninklijk Paleis Amsterdam. P.C.E. 4, 13. SDU Publ. Sept 2008
9. van Bommel B (2013) Terugblik op een geslaagd project. De restauratie van het Koninklijk Paleis Amsterdam. KNOB 1:12–23

10. Brandi C (1979) Intorno a due restauri eccezionali: la facciata di San Petronio e la "Santa Cecilia" di Rafaello. 'Bologna incontri' X(10), 24–27
11. Heinemann HA, van Hees RPJ, Nijland TG, Zijlstra H (2010) The challenge of a perpetual service life: conservation of concrete heritage. In: van Breugel K, Ye G, Yuan Y (eds) Proceedings of 2nd international symposium on service life design for infrastructures. RILEM Publications SARL, pp 1067–1074
12. Symposium 'Growing old gracefully' (Gracieus verouderen), RDMZ, Nov 1999

The Noble Patina of Age

W. J. Quist and A. J. van Bommel

Abstract This contribution discusses the aesthetic aspects of ageing and focusses on the use of the terms 'patina' and 'damage' to decide on the cleaning of historical facades and the application of artificial ageing. Conservation campaigns can be characterised by the wish to preserve an object, building or building complex as a coherent piece of cultural heritage. This contribution discusses the professional debate on balancing between preserving values, i.e. represented by patina and the need to intervene from a technical point of view. Cases of cleaning of limestone and sandstone, together with replacement of natural stone and the application of artificial ageing, are used to illustrate the debate. The perception of professionals is compared with the perception of laymen.

Keywords Building conservation · Cultural heritage · Patina
Damage · Aesthetics

1 Introduction

In daily practice and in most fields of science, ageing is seen as a negative aspect, because of the gradual decrease of the properties of the base material. In the field of architecture, buildings suffer from all kinds of ageing phenomena of which the weathering of exposed materials and the wear of interior materials are the most visible ones. In preservation of cultural heritage, be it either tangible or intangible, ageing, is often referred to as a positive aspect because it makes history visible and therefore plays an important role in decisions about preservation.

W. J. Quist (✉) · A. J. van Bommel
Department of Architectural Engineering + Technology, Section Heritage & Architecture,
Delft University of Technology, Delft, The Netherlands
e-mail: W.J.Quist@tudelft.nl

A. J. van Bommel
Atelier Rijksbouwmeester, The Hague, The Netherlands

© Springer International Publishing AG 2018
K. van Breugel et al. (eds.), *The Ageing of Materials and Structures*,
https://doi.org/10.1007/978-3-319-70194-3_12

153

The wish to strive for preservation of authenticity, visual unity and technical functionality leads to a discussion on the value of patina compared to the trouble of damage, both related to the architectural design and the question of what to preserve.

Decisions about the conservation of historical buildings often depend on technical considerations, but also arguments regarding the artistic value of the object, sometimes including the need for reconstruction of a long-absent structure, and, in practice, non-technical arguments relating to public appreciation, tourist concerns, or even political purposes are taken into account [1, 2]. This paper explores the thin line between 'patina' and 'damage' and how both concepts are dealt with in a technical and esthetical way.

2 Terminology

2.1 Historical Value

Ageing of building materials implies a contradiction. On the one hand, materials are degrading over time and become less functional and less attractive. On the other hand, the ageing of buildings in general and building materials in particular can lead to an increase of 'values' due to an addressed cultural significance and historical importance. Since the mid-nineteenth century, when people became—in a romantic way—interested in the past, we started to preserve historical buildings. Several nineteenth-century scholars, such as John Ruskin (1819–1900), William Morris (1834–1896) and Alois Riegl (1858–1905) tried, all in their own way, to describe what the essence was of historical buildings and why (how) those buildings had to be preserved [3]. Ruskin, for example, writes in his *Lamp of Memory*: "… some mysterious suggestion of what it had been, and of what it had lost; some sweetness in the gentle lines which rain and sun had wrought" [4]. The *Manifesto* by Morris, published in 1877, was a clear pamphlet against the common nineteenth-century behaviour of restoring buildings to the way they might have looked before, without paying attention to the visible traces of time [5]. Riegl, from an art historical point of view, brought 'Alteswert', the value of age, into the discussion [6].

Although people and society have changed over the years, the sentiments regarding the value and meaning of old buildings in general, and built cultural heritage, in particular, have not changed much. The age of a building is until now present in many official documents concerning the formal protection of historical buildings, such as the Dutch Law on Monuments [7].

2.2   Nature of Materials

How does a building become valued in a historical way and when will visible signs of ageing be accepted? When a building has just been finished, damages to materials and constructions are not accepted, and even damages or failures that occur within the first decennium are not accepted. The antagonistic articles by Hendriks in the Dutch periodical *Detail in Architectuur* painfully illustrated damages to modern materials and constructions in recent buildings due to badly designed or executed details [8].

   People are willing to accept a certain degree of degradation due to the ageing of materials, when historical buildings are concerned, or even tend to value buildings or building parts more highly when the traces of time are visible. Research by Andrew [9, 10] indicated that blackening of sandstone facades sometimes added to the appreciation of buildings. Research on the perception of small-scale damage and repairs of natural stone [1, 2] confirms this hypothesis.

   When do buildings start to be liked for their age value and become 'monumental,' how long does it take before people start accepting damages and does it have any relation with the nature of materials? The nature of materials and how people perceive materials probably has great influence on the visible signs of ageing that are accepted. Studies from the 1970s indicated already that concrete is perceived negatively when compared to, i.e. brick or wood [11–14]. The capability of materials to age graciously is probably the most important factor for an old building to become liked [e.g. 15]. Traditional building materials such as natural stone, brick and wood have a certain robustness and are capable of withstanding 'the tooth of time' and therefore are liked. Materials that could be called modern from an architectural point of view such as concrete have an image problem regarding ageing. Most people tend to dislike concrete structures because of their greyish image, illustrated by the debate on the appointment of modern buildings on a heritage list. The qualities of concrete structures, from an engineering point of view and from an esthetical point of view, are often not recognised [16].

2.3   Damage

Assuming that considerations on preservation of historical buildings are often made by persons with various different interpretative frameworks, depending on their education and professional experience, a clear definition of damage would be helpful, as a starting point for any decision on intervention. The definition of damage should be objective and commonly accepted. Several glossaries, combined with damage atlases, have been developed over the years, involving a wide range of specialists in conservation [e.g. 17–21]. Still the definition of damage is rather subjective. Damages to buildings that are technically identical are often handled in different ways, depending on the building concerned and the people involved in the conservation process. Apparently, not only technical arguments are decisive.

Research in The Netherlands and Belgium on the perception of interventions in historical buildings indicated that situations that do not clearly reveal an intervention are appreciated more highly. Whether such situations were 'authentic' or not did not seem to have any influence on the outcome [22]. The harmony or esthetical compatibility of old and new building materials in historical buildings is highly appreciated. Although the parameters are difficult to define and most probable differ according to the cultural and professional background of people, it is clear that ageing phenomena can contribute in a positive way to the valuation of historical buildings. A problem in this discussion however is that, from a scientific point of view, one cannot draw a conclusion by just counting the arguments of individuals. One could say that the majority of people like buildings that look 'authentic', but not draw the conclusion that it is therefore the best to strive for such an image. That would be a simplification in reasoning.

2.4 Patina

The Oxford Dictionaries define patina as 'a green or brown film on the surface of bronze or similar metals, produced by oxidation over a long period,' but, in the context of cultural heritage, patina is often referred to as the visible traces of time on the surface of a material or object [e.g. 3, 15]. The term 'patina' is used to distinguish between the often highly-valued (and to be preserved) traces of time and the undesirable (and to be repaired) 'damages'. This use of the term 'patina' goes back to the 1950s and 60s when there was an intensive scholarly debate among art historians about the cleaning and restoration of paintings, i.e. published in a series of articles in the Burlington Magazine [23]. The famous Austrian-Hungarian Ernst Gombrich (1909–2001) and the Italian Cesare Brandi (1906–1988) took active part in the discussion. The discussion was about the value of patina, which on the one hand shows the age of the painting but on the other hand, due to darkening in time, hides the bright original colours. Gombrich, who was not in favour of cleaning, also brought into the discussion that some painters already took into account the ageing of the varnish by colouring it when applying and, by doing that, anticipating the 'patina'. A comparable example, discovered by Pier Terwen (commissioned by the Dutch Government Building Service, Rijksgebouwendienst), is known from the field of preservation of built heritage: the marble of the monument for Maarten Harpertsz Tromp, in the Old Church in Delft, was originally patinated with a grey paint. Later restorers did not recognise this as authentic and removed almost all the grey paint.

The discussion on patina in the field of conservation of built cultural heritage has always been polarised. On the one hand, the romantic scholars and architects who argue in favour of preserving as much authentic substance as possible and on the other hand the ones that plea—either from a theoretical or pragmatic point of

Fig. 1 Although technically not necessary, visible traces of ageing are often removed, i.e. when cleaning brickwork facades (photo: W.J. Quist)

view—for restoration and reconstruction of the historical architecture. How the (Dutch) layman exactly fits in is not studied, but it is assumed that—looking at many neatly cleaned and restored houses—the majority of the Dutch fit into the second category (see Fig. 1). The boost of facade cleaning in the Netherlands went hand in hand with the renovation of late-nineteenth and early twentieth-century dwellings. Facade cleaning was, first and for all, part of the process in which old dwellings were upgraded to a level that was acceptable by modern standards, and therefore renovated dwellings had to look neat and clean.

3 Decisions on Intervention

3.1 Man-Made Traces Versus Natural Weathering

Article 11 of the Venice Charter states that "The valid contributions of all periods to the building of a monument must be respected, since unity of style is not the aim of a restoration. When a building includes the superimposed work of different periods, the revealing of the underlying state can only be justified in exceptional circumstances and when what is removed is of little interest and the material which is brought to light is of great historical, archaeological or aesthetic value, and its state of preservation good enough to justify the action. Evaluation of the importance of the elements involved and the decision as to what may be destroyed cannot rest solely on

the individual in charge of the work" [24]. This article is often referred to in many
ways to advocate the preservation of the 'architectural layers' in a historical building,
but can it be used in favour of preserving a 'patina'? Figures 2 and 3 show two
different 'traces of time'. Figure 2 shows carvings, used to sharpen a knife or to
gather stone powder with addressed healing capacities, can be called 'mechanical
damage', but, due to their historical meaning, the stone will never be repaired or
replaced. Figure 3 shows a black crust on sandy limestone. Is this 'damage' as safe
as the first one? Although it is a comparable trace of time, the latter will often be
cleaned away, based on a hazy mix of technical and aesthetical reasons.

3.2 Complete Cleaning of Facades

The weathering of (sandy) limestone was studied intensively in the 1980s and 90s
(for example, in the case of the Church of Our Lady in Breda [25, 26]). The general
conclusion about the damaging mechanism that produced the black gypsum crust
was that this process was stable in many places, but it was also concluded that the
dense and stiff cement repointing (from an earlier conservation campaign) had a
large influence on the hygric behaviour of the façade, leading to loss of material

Fig. 3 Black crust due to weathering at Old Church Delft (photo: W.J. Quist)

around this pointing [27]. The architect of the conservation campaign in the 1990s —who was in favour of the complete cleaning of the facades—used the problems of the cement pointing as a technical argument for cleaning: complete removal of the gypsum crust would reveal all damages due to the repointing and would enable the replacement of the repointing and repair of the stone. The necessity of complete cleaning was disputed by several experts, but in the end, the hasty mix of technical and aesthetical arguments resulted in the cleaning [28]. On the one hand, this conservation campaign resulted in loss of the 'traces of time', but, on the other hand, the church was made ready for a new chapter in its long history.

3.3 Artificial Patina

Although only recently applied on several important historic Dutch buildings, the application of an artificial patina on restored facades has a long history. Among (stone)masons it was common to use dirt from gutters or ink to darken newly constructed masonry or pointing next to old masonry [29], but since a few decades

Fig. 4 Application of an artificial patina on new sandstone to visually match old sandstone at St. Lawrence Church Rotterdam (photo: W.J. Quist)

the use of artificial products to colour came in use. Together with careful partial cleaning of old blocks of sandstone, the application of an artificial patina on the brightly yellowish-looking new blocks of Bentheim sandstone helped to unify the architecture of the façade of the Royal Palace in Amsterdam [30–34]. Figures 4 and 5 show the tower of the St. Lawrence Church in Rotterdam, which has recently been restored in sandstone. Partially, the new blocks have been aged artificially to aesthetically match the greyish old stones. The new sandstone for the balustrade and the pinnacles on the corners of the towers were seen as an architectural entity on its own and therefore not aged artificially.

The technical possibilities of applying artificial ageing (i.e. with pigments fixated by ethyl silicate) imposes an ethical question. Can it also be used to colour a different type of stone to match its surroundings? Figures 6 and 7 show two examples of types of stone that were coloured to match their surroundings, not by anticipating its change of colour when ageing, but by altering its natural colour. Figure 6 shows a pinnacle in Peperino Duro coloured as it were Weibern tuff stone. A highly durable type of stone has been used and afterwards coloured to match. Is it fake, or is it an example of 'making use of all possibilities there are'?

Figure 7 shows the effect of ageing of artificial ageing. Due to an almost complete ban on the use of sandstone from 1954 onwards, it was in many cases necessary to choose another type of stone for replacement of degraded sandstone [28, 29]. In the case of the Central Station of Amsterdam, white limestone with an

Fig. 5 Detail of the artificial patina on new sandstone at St. Lawrence Church Rotterdam (photo: T.G. Nijland)

artificial colouring was chosen to replace brownish sandstone. After 25 years the colour on the limestone has been washed away, and both types of stone can easily be distinguished. This example clearly shows that aesthetics were once a reason to apply colour, but is this treatment repeatable, both technically and financially?

4   The Essence of the Debate

An object with historical (cultural) value only has such a value because people impose it on the object concerned. The problem in this statement, however, is that not all people agree on those values. One cannot rely on the majority, and we have to recognise the rights of individuals and minorities. An object with cultural value, first and for all, tells stories that are important to people because these stories are the vehicle of the details of who they are and why they are as they are. That can be a story about the material witnessing of the past, but that is only one of the many stories objects of cultural-heritage value 'tell'. Is an 'objective' outcome of such a discussion about the many historical values of an object thinkable? At least in a lot of cases not, because the discussion on authenticity and on value only shows that both concepts are layered. There is not one form of authenticity (as the Nara

Fig. 6 Colour on Peperino Duro to visually match Weiberner tuff stone at St. Johns' Cathedral, 's-Hertogenbosch (photo: W. J. Quist)

document [35] shows), but there are many forms. As a consequence there also is not one value or not one group of values that can be attributed to an object. The values depend on the point of view of the one who attributes them—subjectivity is unavoidable. If it comes to patina, some will not like it, because patina is disturbing their view of the object. Others however will like the patina; because it shows that the building is old, and therefore the patina is important for the story the building is 'telling' them. Science cannot be the judge in such a debate. The only thing science can do is to look at the arguments that are used. The essay written before the conservation of the facades of the Royal Palace Amsterdam [34] tried to find an answer for this complex debate. The simplified outcome was that there were two main points of view that had to be taken into account. One was the building as one of the most important architectural products of the Dutch Golden Age that had to be 'readable' in its architecture. Ageing disturbed this 'readability'. Some facade cleaning therefore was inevitable. On the other hand, it was important to guard the character of the building as an old building. The approach in this case was to find an intermediate that was acceptable from both points of view. This however does not guarantee that similar solutions can be found in other cases.

Fig. 7 The applied artificial patina on limestone at the Central Station of Amsterdam has been washed away by rain (photo: W.J. Quist)

5   Conclusion

The past decades' theory of architectural conservation has brought forward several concepts on how to handle cases of historical buildings. The most important and well known ones are 'reversibility' as a follow up on the Charter of Venice and 'minimal intervention' as advocated by Brandi [28]. Since the 1990s, 'compatibility' became more and more the leading principle for decisions on intervention in the existing fabric. Teutonico et al. defined it as: "A treated material should have mechanical, physical and chemical compatibility with the untreated historic materials under consideration. Simply stated, compatibility means that introduced treatment materials will not have negative consequences" [36]. This definition is workable, but represents a technocratic view of materials and is missing any link to the aesthetical representation of those materials and structures. Any 'aged' situation can be evaluated according to uniform technical standards, but, when it comes to evaluating the 'visible traces of time on the surface of a material or object', it often comes down to the individual perception of the people in charge. Therefore, we would like to plea for the development of a model for intervention or non-intervention that takes aesthetical aspects into consideration. To make this model and set the criteria, it is necessary to study in-depth how the ageing of materials and structures is perceived by professionals in the field of conservation technology, but also by laymen. Based on the fragmentary research presented in this paper, it is suggested that this could lead to less, but precisely targeted, intervention. This would make the 'patina' not only the result of neglect and undesired ageing, but also the result of careful technical, ethical and aesthetical considerations and therefore truly noble.

References

1. Quist WJ, Van Hees RPJ, Naldini S, Nijland TG (2007) De beleving van schade en reparaties aan natuursteen. Praktijkboek Instandhouding Monumenten (30):15
2. Quist WJ, Van Hees RPJ, Naldini S, Nijland TG (2008) The perception of small scale damage and repairs of natural stone. In: Proceedings of the 11th international conference on durability of materials and components, Istanbul, paper T15
3. Jokilehto J (1999) A history of architectural conservation. Oxford
4. Ruskin J (1989) The seven lamps of architecture, 1849. Dover Publications, Mineola (Reprint)
5. Morris W (1877) The manifesto of the Society for the Protection of Ancient Buildings. http://www.spab.org.uk/what-is-spab-/the-manifesto/. Accessed 23 Feb 2014
6. Iversen M (1993) Alois Riegl: art history and theory. The MIT Press, Cambridge, MA
7. Monumentenwet (1988) http://wetten.overheid.nl/BWBR0004471/geldigheidsdatum_25-02-2014 . Accessed 23 Feb 2014. Since July 1st 2016 the 'Monumentenwet' is part of the 'Erfgoedwet'
8. Hendriks N (2001–2004) Several articles, in Detail in Architectuur
9. Andrew C et al (1994) Stonecleaning—a guide for practitioners. Historic Scotland & The Robert Gordon University
10. Andrew C (2002) Perception and aesthetics of weathered stone façades. In: Přikryl R, Viles HA (eds) Understanding and managing stone decay. Karolinum Press, Praag, pp 331–339
11. Van Wegen HBR (1970) Onderzoek naar de belevingswaarde van vier bouwmaterialen met behulp van de semantische differentiaal – techniek. Centrum voor Architectuuronderzoek. TH Delft, Delft
12. De Jonge D (1971) Over de belevingswaarde van enige bouwmaterialen. Centrum voor Architectuuronderzoek. TH Delft, Delft
13. Brunsman P (1976) Beleving van monumenten. Dr. E. Broekmanstichting, Amsterdam
14. Steffen C (1983) De beleving van gevelvervuiling. Centrum voor Architectuuronderzoek. TH Delft, Delft
15. Denslagen W, Querido J, De Vries A (1978) De tand des tijds; The tooth of time, RV bijdrage 07, SDU Uitgeverij
16. Heinemann HA (2013) Historic concrete: from concrete repair to concrete conservation. PhD thesis, TU Delft, Delft
17. Fitzner B, Heinrichs K, Kownatzki R (1995) Weathering forms- classification and mapping, Verwitterungsformen - Klassifizierung und Kartierung. Denkmalpflege und Naturwissenschaft, Natursteinkonservierung 1. Ernst & Sohn, Berlijn, pp 41–88
18. ICOMOS-ISCS (2008) Illustrated glossary on stone deterioration patterns/Glossaire illustré sur les formes d'altération de la pierre. ICOMOS, Champigny

19. Löfvendahl R, Andersson T, Åberg G, Lundberg BA (1994) Natursten i byggnader. Svensk byggnadssten & skadebilder. Riksantikvarieämbetet, Stockholm
20. Naldini S, Van Hees RPJ, Nijland TG (2006) Definitie van schade aan metselwerk. Praktijkboek Instandhouding Monumenten 28(19):19
21. Van Hees RPJ, Naldini S (1995) Masonry damage diagnostic system. Int J Restorat Build Monuments 1:461–473
22. Quist WJ, De Kock T, Nijland TG, Van Hees RPJ, Cnudde V (2014) Conservering van witte steen: verbetering of verspilde moeite? De beleving van interventies in Vlaanderen en Nederland. In: De Clercq H, Quist WJ (eds) Geological survey of Belgium professional paper 2014/1 N 316, pp 5–14
23. Several articles by different authors, in Burlington Magazine, 1949–1967
24. International Charter for the Conservation and Restoration of Monuments and Sites (the Venice Charter 1964). http://www.international.icomos.org/charters/venice_e.pdf. Accessed 23 Feb 2014
25. Naldini S, Van Hees RPJ (1993) Monitoring the decay of monuments—the Church of Our Lady in Breda—Part A. TNO Building and Construction Research, Delft
26. Verhoef LGW, Koopman FWA (1993) Monitoring the decay of monuments—the Church of Our Lady in Breda—Part B, Delft
27. Quist WJ, Van Hees RPJ (2006) De reiniging van de Grote Kerk in Breda tien jaar later. In: Stokroos ML et al (ed) Praktijkboek instandhouding monumenten. Den Haag
28. Quist WJ (2011) Vervanging van witte Belgische steen. Materiaalkeuze bij restauratie. PhD thesis, TU Delft, Delft
29. Quist WJ, Nijland TG, Van Hees RPJ (2013) Replacement of Eocene white sandy limestone in historic buildings. Over 100 years of practice in The Netherlands. Q J Eng Geol Hydrol. https://doi.org/10.1144/qjegh2013-023
30. Van Bommel AJ (2013) Terugblik op een geslaagd project. De restauratie van het Koninklijk Paleis Amsterdam. Bull KNOB 112:68–79
31. Nijland TG (2013) Reinigen en retoucheren. De restauratie van de zandsteengevels van het Koninklijk Paleis Amsterdam. Bull KNOB 112:112–121
32. Van Bommel AJ (2005) Assessment van ingrepen bij vergrijsde gevels. In: Lagrou D, Dreesen R (eds) Belgische natuursteen in historische monumenten en hun vervangproducten bij restauratie in België en Nederland. 1e Vlaams-Nederlandse Natuursteendag, Leuven. VITO, Mol
33. Van Bommel AJ (2009) Uitgangspunten gevelreiniging gereviseerd. Aantekeningen bij het artikel Assessment van ingrepen bij vergrijsde gevels van 2005. In: Nijland TG (ed) Schoonheidsbehandeling of make-over: Hoe gaan we met de monumentenhuid om? Syllabus TNO-NVMz studiedag, Delft, pp 69–92
34. Van den Ende K, Van Bommel AJ. De gevels van het Koninklijk Paleis Amsterdam. Essay over het voorgestelde herstel van de belevingswaarde.'s-Gra-ven-hage: Sdu-Uitgevers. [Praktijkreeks Cultureel Erfgoed, 4]
35. Lemaire R, Stovel H (1994) Nara document on authenticity. In: Lemaire R et al (eds) Nara conference on authenticity [Conférence de Nara sur l'authenticité]. Nara 1-6-XI-1994. Working papers collected by icomos [Documents de travail rassemblés par icomos]. ICOMOS, Paris, pp 118–20
36. Teutonico JM, Charola AE, de Witte E, Grasegger G, Koestler RJ, Laurenzi Tabasso M, Sasse HR, Snethlage R (1997) Group report how can we ensure the responsible and effective use of treatments (cleaning, consolidation, protection)? In: Baer NS, Snethlage R, (eds) Dahlem workshop on saving our architectural heritage: conservation of historic stone structures, Chichester, pp 293–313

The Ageing of the Creation

L. Melaard-Biçaçi, O. Çopuroğlu and E. Schlangen

Abstract Preserving cultural heritage through restoration and conservation is all about understanding and mastering ageing processes of materials. Appliqué glass artefacts, a Dutch artistic phenomenon of the post-war era in the 20th century, rank among the technically most-intriguing monumental works with multifaceted challenges in conservation. Many of the more historical appliqué-glass panels are nowadays in poor condition as a result of adhesive ageing. While durability and reversibility of the bonding is a major topic, holistic solutions are required considering conservation ethics and integrating cultural value and technical demands. The research project uses an early example of appliqué glass by Karel Appel, The Creation, for studying ageing processes and parameters of glass and adhesives and aims at finding new ways of restoring artworks of appliqué glass.

1 Introduction

The scope of the research project discussed in this article is a neglected area of the Dutch cultural heritage; namely appliqué stained glass in post war architecture. Appliqué glass, also referred to as 'free painting' on glass, can be considered a Dutch invention. Since 1951, applied arts on a monumental scale were integrated in many Dutch buildings with a public function [1].

Appliqué glass had a context within the concept of monumental art in the 1930s characterised by the struggle for a constructive and socially inspired art that was meant for the community and which therefore found its natural place in public buildings [1]. The ruins of the Second World War ironically proved to be a powerful impulse for building and monumental arts. The post-war period (ca. 1945–1965) served as a fertile ground for many new techniques. Especially the monumental art gained a new dimension in this period. Once again, a sense of national

L. Melaard-Biçaçi (✉) · O. Çopuroğlu · E. Schlangen
Faculty of Civil Engineering and Geosciences, Materials and Environment,
Delft University of Technology, Delft, The Netherlands
e-mail: l.bicaci@tudelft.nl

© Springer International Publishing AG 2018
K. van Breugel et al. (eds.), *The Ageing of Materials and Structures*,
https://doi.org/10.1007/978-3-319-70194-3_13

unity and social responsibility was the driving power behind the creations of the post-war period [1, 2]. In this way, the monumental art would free the artist from his social isolation. Many artists were committed to the community through the production of works of art in public buildings. To achieve this, associations were founded and collaboration with architects gained in priority. The government and the industry called for measures to promote the visual arts in architecture. By setting 'percentage regulations' the government also involved the artists in the process of post-war recuperation. In 1951, the percentage regulation of 1.5% was introduced for representative government buildings, in particular, those of public function. In 1953, a 1% *percentage regulation* followed for (private) school buildings. This period of prosperity in the monumental arts ended around 1968 [1].

Glass, appliqué technique involves the use of blown or machine, coloured glass fragments bonded to plate glass with an adhesive. Many Dutch artists of the post-war period, for example, Joop van den Broek, Lex Horn, Harry op de Laak, Berend Hendriks and the well-known Karel Appel, applied this technique to their creations. In comparison with stained glass, the appliqué technique provided a great creative freedom for the artists, without the limitations of a lead network and its specific dimensions [1, 2]. Glass appliqué is not just an art from the past. A beautiful and technically complex recent example is an appliqué-glass window by Marc Mulders called 'Moonlight Garden' 2013 (see Fig. 1).

Fig. 1 'Moonlight Garden' after placement (*left*) and detail of enamel painting on glass (*right*) (photo left: Marc Mulders, photo right: Lisya Melaard-Biçaçi)

As time progresses, ageing constitutes a growing threat to the works of art in appliqué-glass. Many of the more historical glass panels mentioned above are nowadays in poor condition and falling apart as a result of ageing of the adhesives being used, which show yellowing, embrittlement and loss of adhesion. Moreover, growing numbers of buildings from the post-war period in the Netherlands are on the list to be demolished [3]. Some of the windows end up in garbage containers while others are stored with a vague possibility for conservation and placement in the future. However, without any urgent study and intervention, much of valuable information and a learning moment would be lost forever.

Conservation and restoration professionals trying to save appliqué-glass windows from the grip of time are confronted with a bewildering number of challenges. It is not only about which techniques to use to clean the glass fragments, remove the decayed adhesives painstakingly and restore the glass windows to their original brilliance, but more fundamentally their work of reconstructing the original works of art consists of avoiding making the very same uninformed mistakes the original artists of the glass-appliqué period made. This means that studying the fundamental mechanism of ageing is paramount.

This paper presents as a central case study The Creation by Karel Appel that integrates discussion on the boundary conditions or parameters, such as ethical considerations, adhesive properties and ageing phenomenon surrounding the conservation and restoration approach of appliqué glass.

2 Restoration and Conservation of Appliqué Glass

Two aspects of the methodological difficulties conservators of appliqué glass encounter stand out clearly:

- The use of many different types of adhesives and their specific ageing properties; adhesive bonding is an equally cardinal issue in the creation and conservation of appliqué glass
- The complex interplay between technical considerations and cultural opinions on the conservation and restoration of works of art i.e. conservation and restoration ethics

2.1 The Use of Adhesives in Appliqué Glass

The adhesive quality and ageing of adhesives within the context of appliqué glass art has not received much attention until now. There has been no adequate documentation of the materials used. Also, no serious attempts have been made to investigate their in-service behaviour after approximately 60 years of natural ageing. The conservators of historical glass compositions in architectural context are confronted with challenges that need more in-depth and methodical research into this multifaceted issue. Well-studied and scientifically grounded sets of choices are to be made. Given the fact that the appliqué technique is still widely being applied, the current research is not only relevant for the preservation of the early works, but also for the production of future works.

2.2   Conservation Ethics: The Future of the Past

The field of conservation and restoration of cultural heritage unites various domains that correlate strongly, that is to say the contextual aspect and material-technical aspect. The object, the materialized 'finger print' of human history, is subject to ageing.

It must be emphasized that the material-technical aspect of conservation is in the service of the context, to be able to tell the story and its specific meaning that one considers important. A conservative intervention is therefore based not only on the material-technical and aesthetic aspects, but also on the contextual ethical considerations. Therefore, the art, historical and cultural value has to be determined first. However, while determining cultural value is fundamental, prior to any action, one is still left with multifaceted challenges in the area of the conservation and restoration of these multidisciplinary creations. Appliqué-glass windows were always designed as an integral part of the total architectural concept. Demolition of their original location is increasingly a part of the reality today, leading to a loss of context. This creates a new stress field between historical values and integration of these values in a meaning full manner in the present. A particular problematic aspect is also the monumental scale of these creations, which makes them often unsuitable as a part of a museum collection or exhibition. The question rises, how should we display them after eventual restoration? How can we present the original meaning and value of these works of art in a new way, far removed from their original context?

Ethical aspects of our restorative interventions are also complex. As it is confronted in the treatment of The Creation (which is further described in Sect. 7), more than 80% of the lost black paint on part of the fragments should be reapplied. The dilemma is the choice of the substitute material. The original paint layer, presumably East Indian ink, has not proved durable and to re-apply this material would be very problematic. The acceptance of the loss of the black paint without any intervention would mean acceptance of a different appearance than was originally meant. A replacement material must be aesthetically acceptable to match the original, while its identification, as a restorative substitute, should be clear at the same time. Some restorers advocate an archaeological approach while others argue that an accurate documentation is the most important issue that enables us to assess the suitability of techniques and materials in a flexible and realistic manner.

These are some of the dilemmas making clear that restoring appliqué glass is not a matter of just applying the correct technical restoration procedures. It calls for a holistic approach, combining both refined restoration techniques and new and innovative ways of creating present-day meaning from our cultural heritage.

3   Research Context

This study of conservation and restoration of appliqué glass in The Netherlands is related to the project started by National Cultural Heritage Agency (RCE) in co-operation with the Government Buildings Agency (RGD), the Netherlands Institute for Art History (RKD) and the Urban Monument Services of The Hague, Rotterdam and Amsterdam. Their aim is generating awareness among the public at large of monumental art of the post-war period and its cultural value.

The national survey set up in 2008 by RCE 'Help Detecting Monumental Art' is meant to be a framework that enables municipalities, churches, and school boards to assess monumental art in their environment and whether it is of special cultural value. A database specifically designed for this purpose on the RCE site enables input and registration by the public.

4   Research Aims

The research project under discussion aims at establishing three things:

1. A comprehensive survey of all relevant appliqué-glass windows in The Netherlands and a systematic description of all their properties and qualities that can be related to the ageing, restoration and conservation aspects of this particular arts technique
2. Establishing a thorough understanding of the ageing process of glass appliqué and its constituting parameters based on the outcomes of the survey
3. Creating a scientifically sound methodological and practical framework for conservation and restoration of appliqué glass.

5   Research Set-Up

The research project will be carried out in accordance with its three main goals and will therefore consist of three phases:

1. Survey
2. Analysis
3. Creation of a methodological framework.

5.1   The Survey

The aim of the survey is a detailed characterisation of appliqué stained glass by study and documentation of the materials and working practices of the artists/glass workshops. Interviews with artists and glass-art firms which were founded in the post-war period and are still active in design and production of appliqué stained glass can be especially valuable in gathering the necessary information.

Another valuable aspect of the interviews is gathering information about locations where appliqué windows are still present or locating stored ones. This enables the possibility of in situ inspections of representative locations, studying their condition in conjunction with environmental factors and function.

Work practices can be also a contributing factor in the ageing of polymers, especially pre-treatment, and also implementation of the application standards is important. By characterisation and classification of the types of work, the requirements to an adhesive system can be defined in a systematic way. This will influence the choice of adhesives to be tested at a later stage of the research project. It is equally important to gather information about adhesives that have been used until the present through the interviews with the artists and the glass studios. All information will be entered into a database especially designed for this purpose.

5.2   Analysis

The second stage of the project is dedicated to investigate the interplay of material and non-material factors in degradation. The works of art are not only threatened by material-technical aspects but also by a lack of knowledge and infrastructure in the area of preservation and management (see Fig. 2).

Fig. 2 Vicious circle of decisive factors that lead to a decay in the current situation

Since the final aim of this research project is to create guidelines for the restoration and conservation of appliqué-glass windows, we first have to understand the parameters that influence the ageing of these objects. As mentioned before, adhesives and their ageing properties are the key factors in this. Some considerations must be made:

- An important fact to be considered is that windows are functional interfaces between indoor and outdoor environments. Climatic differences, the formation of condensation, the effects of daylight, architectural aspects and imposed vibrational modes are just some of the corresponding factors that a bonded surface must endure.
- Beyond the fact that all polymers are subject to degradation, the sustainability of a bond is also dependent to a considerable degree on the proper selection of adhesives, surface cleaning, adhesive bond-line thickness and accurate application technique.

To determine which types of adhesives are eligible for conservation and restoration of appliqué glass windows in a specific situation, we must determine which requirements are essential for an adhesive, used in this specific context. In other words, the choice of an adhesive depends on its specific application. In this stage of the research, a list of representative research objects will be selected using the information from artist and workshop interviews of the survey phase of the project. Systematic descriptions of their condition in a database will make it possible to study in-service behaviour of the polymers and trace trends in ageing. Collected samples of adhesives will be identified with Fourier transform infrared spectroscopy (FTIR). When sample taking is not possible, Raman spectroscopy will be used.

Accelerated ageing tests will be carried out in conditions of controlled temperature, humidity and UV radiation. Before and after the accelerated ageing, mechanical tests will be carried out. The adhesives for testing will be selected on the basis of the requirements set. Some of the features that must be considered in this context, besides durability, are classified as seen in Fig. 3.

5.3 Creating a Restoration and Conservation Methodology Framework

The third phase of the project is dedicated to the development of a conservation strategy of appliqué glass that integrates the preservation of heritage values and technical demands.

The results of the analysis phase of the research project will provide insight both into the ageing process of appliqué glass and in the feasibility of conservation and restoration possibilities. Given that conservation of appliqué windows almost always involves the total removal of the degraded polymer(s), their reversibility will

Visual properties

- transparent and colourless
- Refractive Index (RI) suitable for glass - Depends on the bonding technique, when application of an adhesive involves the entire surface of the fragment, RI matching is not a cardinal problem. RI matching becomes an important issue when selective area bonding is required

Physico-chemical, mechanical properties

- Adhesion, cohesion - vertical bearing capacitiy of the fragments
- The ability to bridge differences in expantion coefficients
- No, or minimal, shrinkage
- Acid free
- Permits great fluctuations in temperature
- UV-light resistant

Workability

- Allows long processing time - two hours or more
- No air bubbles during processing
- Available, easy to order

In-service behavior

- No yellowing within 30 years
- Reversibility over the course of time
- Maintaining transparency
- No further shrinkage
- Retaining adhesive / cohesive properties

Fig. 3 Considerations for adhesive choice

be a central issue for their conservation. Presumably not all polymers are reversible (or reversible in a convenient way) and alternative choices in such cases are important issues to be discussed and elucidated.

6   The Central Case Study: The Creation by Karel Appel

Development of a conservation strategy will be realized (among others) through the treatment of the central case study The Creation by Karel Appel (1958). Alongside the treatment of the windows, climatic measures in the church, inventory of the aspects for passive conservation and strategies for good maintenance will be formulated.

Karel Appel (1921–2006) is one of the best-known Dutch artists of the 20th century. He also has a considerable monumental oeuvre to his name, which is much less known. Apple made wall paintings, stained glass, tiles, tapestries and carvings. He also designed decors for ballet performances. Through his international fame as a Cobra artist, Appel also regularly received monumental assignments abroad.

The Creation is the first design of six monumental windows that have been realised in this technique for the Paaskerk Zaandam in 1958, commissioned by the architect Karel Sijmons [1].

The modern abstraction of The Creation however was not understood, let alone appreciated. 'Are these the birds? Are these Adam and Eve? It looks like nothing!'[1] were commonly heard remarks from spectators.

Apple's design consists of six windows that depict the story of the creation. The windows have a horizontal oblong design (dimensions 1.12 m × 4.30 m). The sixth window is much smaller in size (1.12 m × 1.57 m). The windowsills run obliquely downwards and contain hot-air diffusers with the opening of the blades into the space. The fragments used to realise Appel's design are variously, blown glass and machine-produced coloured glass. The plate glass carrying the fragments is 8-mm-thick, presumably, drawn glass. However, one window is only 6 mm thick. All six windows contain some fragments that are painted in black with a matt finish (cold paint on mostly transparent colourless glass and some on blue glass). The windows were installed without a buffer layer between the glass and the rabbet. Various shades of blue, red, green, yellow, orange, grey and brown create a colourful vivid image reflected into the space of the church (see Fig. 4).

The condition of the windows was not optimal from the very beginning. Already one year after the windows were placed, the fragments started to detach. Photographs from 1959 already show areas of discoloration of the adhesive used.

In 2004, Cultural Heritage Agency of the Netherlands (RCE) prepared a condition report on the windows. The report points out the poor condition of the windows, such as rotten window frames and the loss of the majority of the black paint.

In 2005, the Reformed Church (the owner at this time) sold the Paaskerk to the Free Evangelical Church of Zaandam. The appliqué-glass windows were sold for a symbolic price of one euro.

In 2008, the church was given the status of a monument. In 2012, sufficient funds were collected for the conservation project of the windows.

Eight years after the first inspection, the window frames proved to be badly rotten, and also about more than 80% of the black paint had been lost (see Fig. 5 right). Many glass fragments had come loose and were kept in boxes by the church community. Part of the fragments had fallen into pieces. Furthermore, a greasy grey veil of dirt, algae and a collection of insects in the openings left between the loosening fragments and the plate glass were present (see Fig. 5 left). The windows had not been cleaned for more than 10 years, as cleaning such a complex surface was not an easy task. The adhesive that presumably had been applied over the whole surface of the fragments had disintegrated from the heat and slipped down to the lower planes of the windows, leaving some of the glass fragments fixed at one point only.

---

[1]Oral delivery by one of the spectators A. Engels, 2013.

Fig. 4 From top to bottom; day one: separation between light and darkness; day two: separation between the air and water; day three: creation of earth and vegetation; day four: creation of the sun moon and stars; day five: creation of marine life and air live; day six: creation of terrestrial animals and Adam and Eve, the church symbolizes the seventh day: rest (photos: Lisya Melaard-Biçaçi)

Fig. 5 Disintegration and migration of the adhesive to the lower planes (yellow fragment photo *left*), algae and fungi growth (bottom of photo *left*), loss of black paint around the coloured fragments now seen as colourless, transparent fragments (photo *right*) (photos: Lisya Melaard-Biçaçi)

The adhesive was significantly yellowed and brittle. Two of the windows were damaged by vandalism and showed large diagonal cracks.

7 Preliminary Research Result

The research project discussed in this article started in September 2013 with the systematic survey of applique-glass objects in the Netherlands and the first phase of the restoration of The Creation. Some preliminary findings can already be mentioned here:

1. There are considerably more applique-glass objects in The Netherlands than recorded in the official registers of RCE. The official registers are contributed by a wide public and therefore contain some mistakes. Some of the artwork is attributed to the wrong artist, or the record has incorrect dimensions. Furthermore, quite a few objects are registered erroneously more than once.
2. Not just one or two, but a wide and quite heterogeneous variety of adhesive types, were used in making applique glass works of art, complicating the analysis of ageing parameters significantly.
3. First research on The Creation shows that applique glass objects are far moré complex in structure and properties than appears at first glance.

7.1 Recordings of Applique Glass Objects

The database of the RCE contains approximately 52 windows. Artist and workshop interviews conducted for this research project resulted in 22 more objects that are

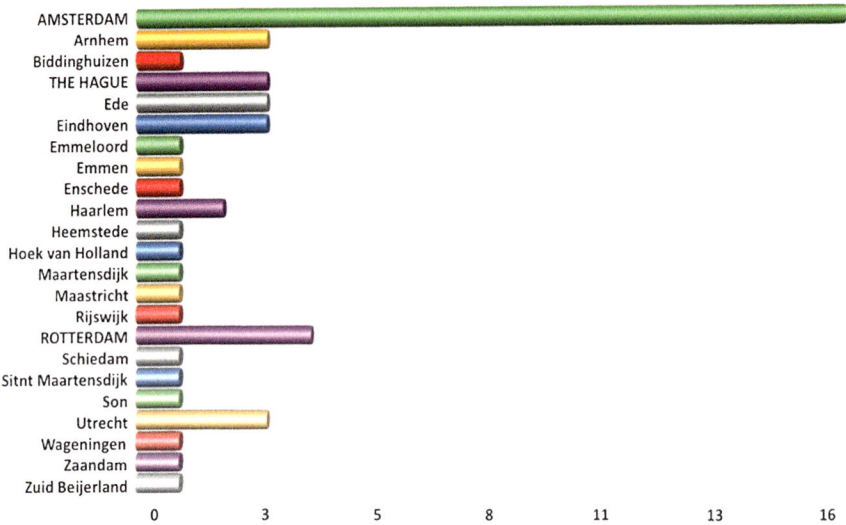

Fig. 6 Geographical distribution of 52 applique-glass works of art

not recorded in the RCE database. Possibly yet more windows will be found and recorded during this project. The geographical distribution of recorded applique glass objects in 23 locations' in the Netherlands is presented in Fig. 6.

7.2 Applied Adhesives

First survey results show that, in The Netherlands epoxy, acrylic, polyurethane, polyvinyl acetate, UV-hardening acrylates and silicone polymers have been utilised for more than a hundred examples of applique stained glass from the second half of´ the twentieth century (see Table 1).

7.3 The Central Case Study—The First Stage of the Treatment

In the first restoration phase prior to the removal of the windows, a comprehensive documentation was executed. A life-size drawing of all the windows, with fragment numbers and detailed pictures was made, marking missing fragments and areas showing loss of paint.

The somewhat loose shards were temporarily attached with a wide adhesive tape. The windows with large diagonal cracks were reinforced to prevent breakage

Table 1 First survey results, information gained from the artist and glass studio interviews

Adhesive	Type of polymer	Remarks by glass studios
Acrifix 90 Rohn & Haas (with primer)	Two comp. acrylics	Used from ca. 1965 until the 1980s
Acrifix190 Rohn & Haas (With primer)		Relatively flexible adhesive that was suitable for a multi-layer appliqué glass design
		Production was ended in the 1980s due to environmental measures
		Slight yellowing has been observed. However, it is noted that application norms have not always been followed accurately
Acrifix 93 Rohn & Haas (with primer)		
Bohle Verifix MV760VIS	Acrylic with UV-hardener	By glass workshops defined as suitable for small-scale windows due to high prices and rigidity
Bohle Verifix MV760		
Bohle Verifix LV760VIS		
Araldite CY221/harder Aradur HY2967	Two comp. epoxy	In use ca. from the 1980s until the present
		Yellowing and embrittlement
Araldite 220		(indoors, less problematic)
Araldite 2020		Additives such as primers, UV-filters and plasticisers were (are) in use by some studios
Araldite HA259		Difficulties in reversibility
Araldite HA258	Three comp. epoxy	
QX sil12	Two comp. silicon	In use since ca. ten years
		Some comment that it withstands humidity but has weak bonding properties
Sil Gel Wacker Harder 612B		
RTV30 General Electric		Weak bonding, requires specific climate; glass studios often do not have the facilities

during disassembly. In total, approximately 1,500 fragments were removed and cleaned by use of chemical and mechanical methods, using alternately hot water and/or ethanol (see Fig. 7). All fragments with black paint will be treated separately in later stage to avoid direct contact with water or solvents by the use of poultices.

Identification of the adhesives that had been used was carried out by RCE.

The results of FTIR spectroscopy show that polyvinyl-acetate adhesive has been used for the creation of the windows (see Fig. 8). However, some fragments were obviously bonded using silicon adhesive. Presumably loosened fragments were bonded with silicon adhesive over the course of time.

Although identification of the composition of the black paint has yet to be carried out, artist interviews suggest that most probably East Indian ink was applied. Samples of black paint from various fragments will be tested with gas

Fig. 7 In addition to introduction of ethanol by capillary action, mechanical intervention by a fine steel spatula was necessary to remove the fragments (*left*); fragments were subsequently cleaned by immersion in warm water that proved to be effective (*right*) (photo: Lisya Melaard-Biçaçi)

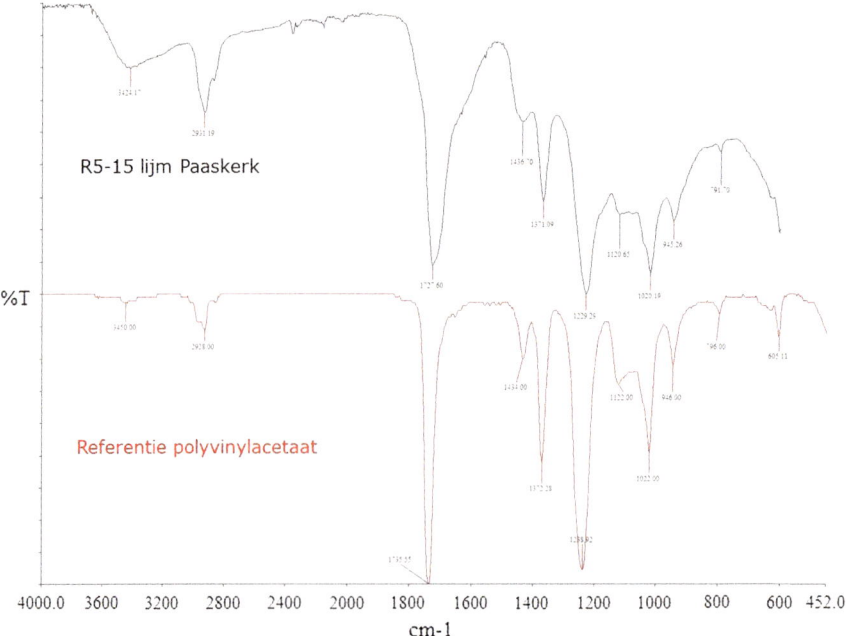

Fig. 8 FTIR analyses were performed, using a Perkin Elmer Spectrum 1000 FTIR spectrometer, combined with a Grase by Gate Single Reflection Diamond ATR. The analysis was carried out by the research department of RCE

chromatography. As mentioned before, more than 80% of the paint layers has been lost, therefore alternative materials will be tested for their adhesion to a glass surface with a matt finish and for durability.

8 Conclusions

The research project has yielded so far the following preliminary conclusions:

1. Detailed analysis of The Creation makes clear that glass applique works of art are far more complex than is apparent at first sight. The structure, weight and dimensions of glass fragments applied in one object (or between different objects) can vary enormously. Classification of types of applique glass is necessary to define requirements for an adhesive system in various cases. In general, all kinds of glass colouring can be found and also a wide variety of bonding materials. This will probably mean that identifying the operative ageing parameters in glass applique will call for far more comparative research, such as the systematic study of in-service behaviour of adhesives and accelerated aging tests than were envisaged at the start of this research project.

2. The first phase of the conservation of The Creation has shown that the restoration of these applique glass works will involve, in many cases, the complete dismantlement of the glass structure and removal of all remains of the adhesives used. This means that reversibility is a central and critical factor in choosing the right conservation materials and techniques. The field survey to be conducted therefore has to focus on the feasibility of restorative interventions that depend on reversibility of the used adhesives after ageing. This is an important issue in testing suitability of new adhesives.

3. While cleaning The Creation, very diverse surface changes on glass have been observed, depending on the type of glass and glass colour. It must be emphasised that glass-to-glass bonding concerning new creations could be a different issue than the bonding of historical glass for conservative intervention. Physicochemical changes on historic glass surfaces might influence the quality of an adhesive bond. This has to be taken into account when formulating the restoration methodology framework intended in this research project.

Acknowledgements Thanks are due to the following for assistance and helpful information in relation to the research project described in this paper: Glass studios; Van Tetterode, Glas Bewerking Bedrijf Brabant (GBB), Bevo Glass Art, Atelier 70, Henk van Kooij. Glass artists; Louis La Rooy, Marc Mulders, Suzan Drummen, Joost van Santen, Norbert van den Broek. National Cultural Heritage Agency (RCE); Rutger Morelissen, Simone Vermaat, Geertje Huisman, Suzanne de Groot, Luc Megens. The authors also gratefully acknowledge the assistance of Jacob Molenaar, the team of Paaskerk Zaandam, conservators Eva Wolfs and Cock Tesselar. Special thanks to Frits Kratz and Marga Hessling for their contribution as a film crew for the artist and glass studio interviews.

References

1. Hoogveld C (1989) Monumentale beglazingen gedurende de periode 1945–1968. In: Glas in Lood in Nederland 1817–1968 1st edn. SDU, 's – Gravenhage, pp 163–195
2. Van Dael P (1989) De Monumentalen. In: Glas in Lood in Nederland 1817–1968 1st edn. SDU, 's – Gravenhage, pp 108–141
3. Abraham B (2011) Vergeten Kunst. Monumentale wandkunst in Nederland 1945–1965. Vormen uit Vuur 214:24–29

Part VI
Modelling of Ageing Materials and Structures

Modeling Ageing Cementitious Pore Structure

Neven Ukrainczyk and Eddie A. B. Koenders

Abstract Coupled reactive-transport processes in cementitious materials play a crucial role in the ageing process of building materials. Up to now, the effect of ageing on microstructure and evolution of their properties, in three dimensions, was studied only by the dissolution of certain phases at random locations. In this paper, we present a new 3D reactive-transport model and the results of leaching-induced ageing simulations performed on virtual cementitious microstructures generated by the Hymostruc model. The outputs are the morphology of the aged-3D microstructure together with aged properties characterised by the fuzzy state of each voxel at different times. This enables simulating the evolution of properties as a function of time, as well as a function of the location within the microstructure. Correlation results obtained from pore-scale (3D) modeling should then be used to make a homogenized, predictive model to be used for the design (ageing assessment) of full-scale structures.

Keywords 3D modelling · Reactive-transport · Cementitious materials
Calcium leaching

1   Introduction

The knowledge of the relationship between microstructure (i.e. pore structure) and transport properties, e.g. effective diffusion coefficient, D_{ef} of cement-based materials is a crucial aspect to generate properly the models coupling transport and chemistry. During the transport mechanism, the local microstructure of the cement-based material is changing due to the dissolution effect and/or interaction of

N. Ukrainczyk (✉) · E. A. B. Koenders
Technische Universität Darmstadt, Franziska-Braun-Straße 3,
64287 Darmstadt, Germany
e-mail: ukrainczyk@wib.tu-darmstadt.de

© Springer International Publishing AG 2018
K. van Breugel et al. (eds.), *The Ageing of Materials and Structures*,
https://doi.org/10.1007/978-3-319-70194-3_14

the diffusing ions with the solid phase. Thus, the coupled-transport and chemical-equilibrium effects result in a space and time dependency of the D_{ef}. Both these dependencies $\{D_{ef} = f(t, \text{x})\}$ present an important backbone of the coupled-modeling concept and show significant impact on the final results of the predictions.

Modeling of calcium leaching of cementitious materials is challenging [1–4] due to the multi-scale porous and multi-phase nature of the cement matrix. The mechanism of calcium loss and the equilibrium-calcium concentration (solubility) are different for each of these phases, depending mainly on their calcium-to-silicon ratio [4]. The most common approaches to overcome these difficulties consist in either limiting the number of phases considered or using a continuous-equilibrium formulation based on Berner's equation [5, 6] for solid calcium concentration as a function of its equilibrium-solution concentration.

The rapid development of numerical models has provided novel methods to investigate the influence of microstructure on the evolution of the properties of cement-based materials. A numerical scheme that reflects a representative porous network can be used to analyse the effective transport properties by means of either generating a porous network via advanced 3D-numerical simulations (e.g. CEM-HYD3D, µic or Hymostruc [7]); or by 3D sampling of a porous network using modern experimental imaging techniques, such as X-ray computed tomography. A virtual-3D microstructure created with an available hydration model provides a basis for the analysis of the morphological influence on the effective diffusion coefficient. Such an approach contributes to a better understanding of the phenomenology and thus improves the predictive reliability of the coupled models.

The effect of ageing on microstructure and evolution of its properties, in three dimensions, was studied only by dissolution of certain phases at random locations. In this paper, we present a novel, 3D, reactive-transport model and the results of leaching-induced ageing simulations performed on virtual cementitious microstructures generated by the Hymostruc model. This enables simulating the evolution of properties as a function of time, as well as a function of location within the microstructure.

2   Homogenised Model

The calcium leaching in cement-based materials is a coupled-chemical equilibrium/diffusion phenomenon [2]. The kinetics of this transport-reaction processes are described by a simplified continuum (homogenized) model Eq. 1 where the actual evolution of the pore morphology, chemical activity, convection, electrical coupling and precipitation effects are neglected. However, as an initial approach, it makes a good prediction of the degraded depth and cumulative amount of leached calcium. The homogenised approach requires empirical relationships linking the diffusion coefficient to porosity:

$$P(x,t)\,\partial u/\partial t = D_{eff}(x,t)\,\partial^2 u(x,t)/\partial x^2 - \partial u^{solid}(x,t)/\partial t \qquad (1)$$

where $u(x, t)$ is the Ca^{2+} concentration in the pore solution, $u^{solid}(x, t)$ is the concentration of Ca in the solid phase, $P(x, t)$ is the porosity and $D_{eff}(x, t)$ is the effective diffusion coefficient of Ca^{2+} ions (790 μm^2 s^{-1} [8]). The first term on the right-hand side of Eq. 1 stands for the diffusion process of the calcium in the liquid phase, which is assumed to be governed by Fick's law. The second term accounts for the dissolution process, which leads to a source of calcium in the liquid phase. The calcium concentration in the solid phase $u^{solid}(x, t)$ is calculated from its relationship with calcium concentration in solution.

The existence of a solid-phase assemblage with clearly defined dissolution fronts [1, 2] is explained by instantaneous dissolution, i.e. establishment of the local-liquid equilibrium concentrations. The kinetics of the dissolution reaction is governed by a diffusion process because the diffusion rates are much slower than those of the chemical reactions. The evolution of Ca/Si ratio as a function of calcium concentration in solution corresponds to the degradation fronts from the non-degraded to the external zone of a cement material. The dissolution of portlandite occurs suddenly for a threshold calcium concentration of 21 mol m^{-3} and explains the rapid drop at this value. The decalcification of CSH is gradual due to the various forms of CSH having a Ca/Si ratio ranging from 1.65 to one. This explains the decreasing calcium concentration in pore solution between 21 and two mol m^{-3}. Dissolution of ettringite and mono-sulphoaluminate also occurs in that zone. For calcium concentrations under two mol m^{-3}, the solid phase corresponds to a silica gel.

3   3D Virtual Microstructure Generation

The 3D-virtual cementitious microstructures generated by Hymostruc can be simulated as a function of the random positioning of the cement particles inside a predefined REV, the initial particle-size distribution (PSD), the degree of hydration, the chemical composition of cement, the morphological development of the hydration products, the water to cement ratio and the temperature of the reaction process. The initial state of the microstructures is determined by stacking cement particles that follow a predefined particle-size distribution (Fig. 1a). For this, periodic boundary conditions were applied to minimise size effects and to comply with the volume balance induced by the water-to-cement ratio. Particles are stacked based on random selection of locations with equal probability of occurrence. Placing of the particles starts with the largest particles followed by smaller ones, while obeying the particle-size distribution. This process continues until all particles in the smallest fractions have been stacked. After having generated this initial particle structure, hydration algorithms are invoked that simulate the stepwise evolution of the particle-hydration process and associated expansion of the outer shells of hydration products, whilst forming a 3D-virtual microstructure (Fig. 1b).

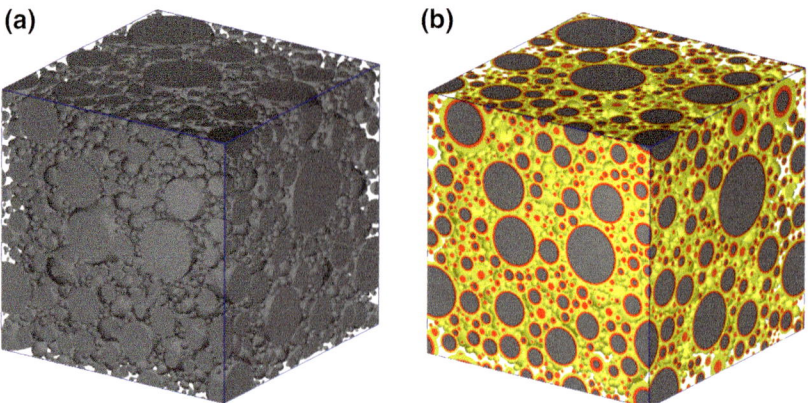

Fig. 1 3D simulated microstructure (grey-cement, red-inner hydration product, yellow-outer hydration product): **a** initial unhydrated cement; **b** after hydration

During this hydration process, the solid volume fractions of the reactants, i.e. the anhydrous cement and free water, decrease, while, in return, the total fraction of formed hydration products increase. The outer expansion of the individual particles is calculated according to the so-called particle-expansion mechanism [9], which accounts for the geometrical expansion of the expanding particles that overlap smaller particles and are located nearby. The positions of the solid particles in space are described by means of a vector in a 3D-Cartesian coordinate system. This includes the start (0, 0, 0) and end location (x, y, z), the diameter of anhydrous cement grain and the thickness of the inner and outer hydration layers that surround the shrinking core of the anhydrous cement particles.

4 Transient Transport Model

A finite-difference (FD)-based numerical scheme is derived for solving the transient transport problem (3D extension of Eq. 1, only for transport). The algorithm starts with a discretization of the 3D-virtual microstructure into a regular 3D mesh (e.g. Fig. 2a), where each voxel in the mesh is assigned to be either a capillary-pore phase or a solid phase, according to its actual position in the microstructure. In this algorithm, identification is done by the center point of a voxel. For each voxel plane that shares with a neighboring voxel in x, y and z direction, a conductivity coefficient c_x, c_y, and c_z, needs to be assigned, respectively (Fig. 2a). The connectivity of all voxels are stored in three **c** vectors (whose lengths correspond to the number of voxels in the system, N). A six-neighbour configuration was used, representing a situation where the voxel is connected to its neighbours by the six planes of a cubic voxel in x, y and z direction. The conductivity coefficients of the surfaces that

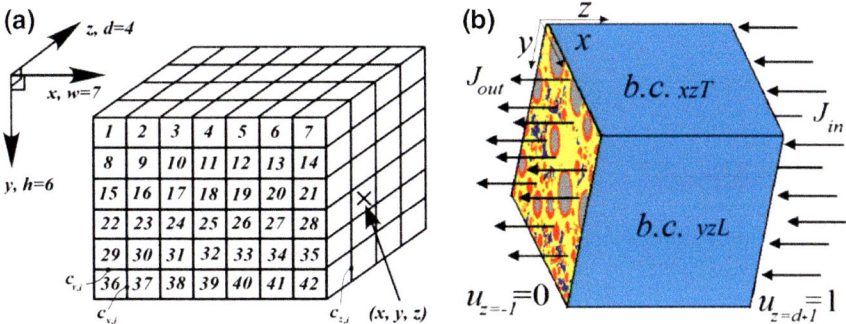

Fig. 2 a FD implementation, position and size of coordinates: width (x), height (y), and depth (z). Each sharing surfaces between neighboring voxels in x, y and z directions has an assigned conductivity coefficient c_x, c_y, and c_z, respectively. **b** Steady-state flux (J) across z-axis with (periodic or non-periodic) boundary conditions (b.c.) employed on the four side faces parallel to the imposed flux

connect a central voxel to its neighbour voxel are calculated from a series-connection approach using two conductors according to Eq. 2:

$$c_i = 1/(0.5\ /D_i + 0.5\ /D_{i+k}) \tag{2}$$

where $k = 1$, w, or L. With this notation, w represents the number of voxels in a row and L the number of voxels in a layer. Fick's second law is solved by a second-order finite-difference scheme provided in Eq. 3. This equation shows a backward Euler fully-implicit form of finite difference, for node i and its six neighbours.

$$c_{x,i-1}u_{i-1} + c_{y,i-w}u_{i-w} + c_{z,i-wh}u_{i-wh} +$$
$$- \left(\frac{\Delta x^2}{\Delta t\,D} + c_{x,i} + c_{x,i-1} + c_{y,i} + c_{y,i-w} + c_{z,i} + c_{z,i-wh} \right) u_i +$$
$$+ c_{x,i}u_{i+1} + c_{y,i}u_{i+w} + c_{z,i}u_{i+wh} = - \frac{u^{OLD}\,\Delta x^2}{\Delta t\,D} \tag{3}$$

where u is the Ca concentration in solution that needs to be calculated for all voxels and for each time increment, based on the known concentration u^{OLD} of the previous time step. Assembling the equations for all (N) FD nodes forms a global system of equations, which can be represented in a matrix notation by Eq. 4:

$$\mathbf{A\,u = b} \tag{4}$$

where u is the voltage vector (size of the total number of voxels in the system, N), $\mathbf{A}$ is a sparse and symmetric matrix with seven diagonals (each voxel has six nearest neighbours) that contain information about conductivity coefficients of all the connections among the voxels and $\mathbf{b}$ is the vector of known concentrations (i.e.

boundary condition and previous time-step concentrations u^{OLD}). The obtained system of Eqs. (3), and (4) is solved by a conjugate-gradient algorithm with an optimised matrix-vector multiplication. This has been achieved by multiplying only those elements of the matrix that lie on the seven diagonals, while avoiding multiplication of a very large number of zero elements. Furthermore, since the size of the sparse matrix $\mathbf{A}$ is N times N, and may reach huge dimensions, the matrix is not stored explicitly but only by means of the vectors $\mathbf{c}_x$, $\mathbf{c}_y$ and $\mathbf{c}_z$, which store the conductivity coefficients of the connections between voxel planes in the x, y and z directions, respectively.

5   Boundary Conditions

The main flow direction of the REV samples simulated in this study is considered to take place in z-direction. The four surfaces that are situated parallel to this main flow direction are subjected to two different types of boundary conditions, i.e. periodic and non-periodic. The non-periodic boundaries represent a full sealing off of the four boundary planes that run parallel to the dominant flow direction of the imposed flux (z-direction, Fig. 2). This means that no transport can take place through these surfaces at all. On the contrary, periodic boundary conditions are also applied to the REV samples under consideration and represent a full disclosure of the four parallel boundary planes, which are connected to the planes situated at opposite sides of the sample.

   In the numerical algorithm, the boundary conditions are handled as two additional coefficient vectors ($\mathbf{p}_x$ and $\mathbf{p}_y$), each with a length equal to the number of boundary voxels that form one surface of the system: e.g. length $= (h - 2) * w$, where h is the height and w the width of the surface. These two vectors store the conductivity coefficients for each element at the boundary surface. If one of the voxels at a surface is a solid (zero conductivity), then the conductivity coefficient of the element it is linking to in the opposite surface plane is zero (no connectivity, and no flux). On the contrary, if a pore element in a surface is connected with a pore element in the opposite surface plane, then a flux is possible in that particular direction. After assembling of the main matrix according to Eq. (6), the boundary conditions are applied.

6   Degradation Code

The numerical implementation is following the transient FD scheme for transient transport (section transport), including the following modifications to incorporate the degradation mechanism and source term for dissolution of Ca. The main idea is to locate the solid voxels that are in contact with the pore solution and impose their solution concentration from the previous 0, corresponding to solid voxel, to a saturation value (e.g. 21 mol m^{-3} for portlandite). These phase-boundary voxels are then dissolved by applying the mass-balance calculation based on the total flux emanating from the corresponding voxel to the nearest neighbours. The total amount of Ca diffused into solution (i.e. neighbouring voxels) per time increment is equal to the amount of dissolved solid Ca and is calculated by Eq. 5:

$$\Delta n = D\,\Delta t\,\Delta x^2 \sum_{i=1}^{6} \Delta u_i/\Delta x_i \qquad (5)$$

The amount of moles required to saturate the pore solution of the dissolving voxel, was not considered in this preliminary calculations, avoiding nonlinearity due to dependency on increasing porosity. By knowing the rate of the dissolution, which was equalised to the diffusion rate, the change of porosity of the dissolving voxel with time can be calculated using the value of the bulk density ($\rho = 2.27$ g cm^{-3} for portlandite) and molar mass (M_r):

$$\Delta P = \Delta n\,M_r/\rho \qquad (6)$$

The dissolution of the Ca(OH)$_2$ occurs until the solid Ca concentration reaches 0 and, after that, the voxel is treated as a (100%) pore voxel. Following the same approach, the change of Ca/Si ratio in the CSH phase can be calculated. For this phase the solubility, i.e. equilibrium concentration of the solution, that represents the threshold value, below which the dissolution occurs, depends on the Ca/Si ratio.

The phase-boundary voxels are found according to the criteria of having at least one 'pore-voxel' neighbour. The FD equation for that voxel and the corresponding neighbours is updated. The pore concentration of the dissolving voxel is forced to the saturation (equilibrium solubility) concentration. The connectivity coefficients c connecting the dissolving voxel to the nearest neighbours are changed from 0 (no connectivity with a solid voxel) to a relative value of 1, corresponding to a pore-pore connection. The FD equations for the neighbours are also adjusted to consider the newly-introduced saturated concentration on the dissolving voxels. This known concentration value becomes the right-hand side, i.e. vector **b**, of Eq. 4.

7 Simulations

Two 3D-virtual microstructures were generated to test the new degradation model: (1) simple sphere and (2) virtual cementitious microstructure. Specimens were cubic with a size of 50 and 1 μm voxel. The cement pastes were prepared with an initial water-to-cement mass ratio of 0.4, hydrated for 28 days, resulting in a 13% capillary porosity.

The model is based on the resolution of the mass-balance equation for calcium concentration in solution and in solid. In this first presentation stage, the whole degradation process was demonstrated solely in terms of portlandite's dissolution. In other words, all of the solid phase was treated as portlandite. Instantaneous dissolution is assumed because the diffusion rates are much slower than those of the chemical reactions, so the kinetics of degradation is governed by a diffusion process. The dissolution of portlandite occurs suddenly below a threshold calcium concentration of 21 mol m^{-3}. The initial condition of the Ca concentration in the capillary pores was put at this saturation value. The aggressive solution is added by specifying Dirichlet boundary conditions (meaning fixing a constant concentration on the boundary to 0), simulating that one side of the model sample is placed in a reservoir of deionized water. This boundary condition is very rigorous, which, in reality, can be obtained by rapid (turbulent) flow of the water through the external reservoir.

8   Results and Discussion

Figure 3 shows 2D slices taken from the middle of the 3D results of the simulation obtained on two initial microstructures: (1) simple sphere (upper sequence in Fig. 3) and (2) virtual cementitious microstructure (lower sequence). The Dirichlet boundary condition was applied on the left side of the slices. The presented degradation times are: 0, 10, 250, 500 and 1000 s. After 10 s of simulation, both initial microstructures have increased in porosity because the phase-boundary solid voxels are transformed to a saturated concentration. With further degradation, the results show a moving front of the concentration gradients that gradually dissolves the solid (Ca(OH)$_2$). The dissolution front is sharp and is located always at the right end of the concentration gradient. Inside the porous matrix (further right from the dissolution front), there is no concentration gradient and therefore no dissolution of the solid. The inclusion of CSH solubility in the model, which has variable solubility with the Ca/Si ratio, is expected to create a more distributed dissolution in the matrix, which is not only limited to the moving front.

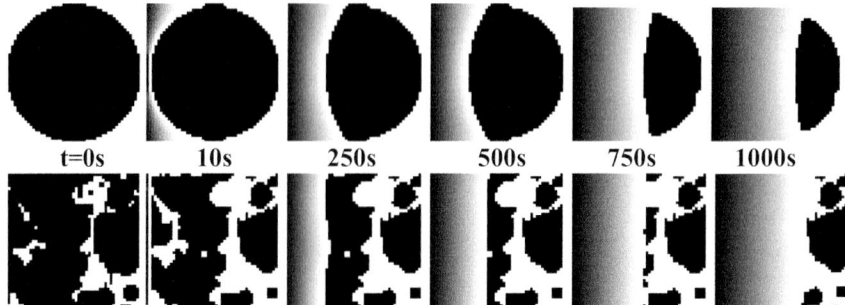

Fig. 3 A 2D slice of the 3D simulation results at degradation times: 0, 10, 250, 500 and 1000 s applied on two initial microstructures: simple sphere (**upper sequence**) and virtual cementitious microstructure (**lower sequence**). Grayscale map represents the concentration distribution (black is 0 and white is 21 mol m^{-3}), while black also represents a solid

Next, the validity of instantaneous dissolution is discussed. Wang et al. [10] measured the dissolution rate of lime in solutions using a rotating disc of compressed lime to control the mechanisms involved in $Ca(OH)_2$ dissolution and the factors that affect the dissolution. The lime dissolution rate varied with the rotational velocity for values below 300 rpm, which is a threshold value for laminar and turbulent flow. When the disk rotational velocity was greater than 300 rpm, the dissolution rate was constant at 5.5 10^{-16} mol $(\mu m^2 \ s)^{-1}$ (at pH = 7 and T = 25 °C). In Portland-cement pastes, the pore solution typically has a pH from 12 to 13, the value depending in part on the concentration of soluble alkali salts in the cement. The dissolution rate decreased exponentially with increased values of pH over the pH range 3–7, therefore, a value lower than 5.5 $\times$ 10^{-16} mol $(\mu m^2 \ s)^{-1}$ is expected in the cementitious materials. On the other hand, the higher the surface area for a given portlandite mass, the higher the dissolution rate. Considering the surface area of 45 $\mu m^2/\mu m^3$ for CH [11] would yield a 45-times-higher dissolution rate, i.e. 25 10^{-15} mol $(\mu m^2 \ s)^{-1}$. This rate is still slightly higher than the maximal flux scenario occurring for the first time steps at the Dirichlet boundary condition, where a 0 concentration is directly connected to a saturated concentration of the dissolving voxel, calculated as flux = $D \ du/dx$ = 790 $\mu m^2 \ s^{-1}$ 21 10^{-18} mol μm^{-4} = 16 $\times$ 10^{-15} mol $(\mu m^2 \ s)^{-1}$. Therefore, care should be taken to check the validity of the assumption that the diffusion rate is the rate-controlling mechanism. If the diffusion rate is higher than the dissolution rate, the concentration of the dissolving voxel goes below the saturation value. In the proposed degradation algorithm, this scenario is implemented by comparing the total flux from the dissolving voxel with the user-prescribed dissolution rate (which depends on the temperature, pH and surface area), and, if the flux is greater, the concentration of the voxel is lowered according to the amount of excess flux, porosity and volume of the voxel.

9 Ageing Phenomena in Cement-Based Materials, Linking the Issues of Ca Leaching to Relevant Durability Problems

Ca leaching is of great importance in the utilisation of concrete as a solidifying barrier for the geological disposal of radioactive waste. Here, the design requires lifetime of thousand years, which far exceeds the durability of regular concrete structures. The issues of Ca leaching may be of concern for the durability of concrete. For example, CH dissolution increases the porosity of the surface layers of concrete and detrimentally affect the resistance of the material to de-icer-salt scaling and ion penetration. In other examples, the leaching of calcium may also affect the bulk of the material and have a detrimental effect on the mechanical and transport properties of cement-based materials. The service lives of structures are often evaluated through indicators like porosity, diffusivity and water or gas permeability. In water conduits in connection with a hydro-electric power scheme, it was found that the surface of the concrete had been attacked and the aggregate had been exposed within a few year's exposure to flowing waters. After a period of years, the coarse aggregate near the surface became loose and commenced to be washed out.

10   Link of Model–Structural Performance in the Long Term with Respect to Ageing of Materials and Structures

Ca leaching may change cement-based materials' pore microstructure, mainly due to portlandite dissolution. By this dissolution, the more porous microstructure influences the long-term performance of cement-based materials and structures, especially when exposed to pure water and flowing streams. The prediction of Ca leaching over extremely long periods, e.g. for nuclear-waste containments, requires combining of (accelerated) experimental methods with numerical simulations for more rational evaluation of the service life.

In future works, the multi-scale modeling should contain three length scales: (i) nano-porosity model for C-S-H (e.g. by Monte Carlo protocols); (ii) capillary pore scale (Hymostruc) model; and (iii) meso-scales originating from aggregates cement–paste interactions, explicitly considering porous interfacial effects. The reactive-transport model should consider: (a) electrical coupling between multi-component diffusive ions; (b) time and space dependency of the parameters; and (c) chemical interaction of the diffusing ions with the solid phase (e.g. binding and dissolution effects), water movement (capillary suction) effect and migration of ions under the externally-imposed potential difference. Correlation results obtained from pore-scale (3D) modeling should then be used to make a homogenised predictive model to be used for the design (ageing assessment) of full-scale structures. The proposed complementary modeling approach will represent a powerful new tool to couple reactive-transport in cementitious materials with the effect of the ageing process at structural level. The consequences of calcium leaching, in terms of mechanical behaviour, have been experimentally studied and show a decrease in strength and a large increase in creep rate. Modeling these phenomena requires combining chemical (reactive transport in porous media) and mechanical modeling.

11   Conclusion

A new 3D reactive-transport model and results of leaching-induced ageing simulations were performed on virtual cementitious microstructures generated by the Hymostruc model. The preliminarily simulation results were demonstrated by considering only the portlandite dissolution.

The outputs of the new approach are the morphology of the aged 3D microstructure together with aged properties characterised by the fuzzy state of each voxel at different times. The degradation simulation results show a sharp moving front followed by the concentration gradients that gradually dissolve the solid ($Ca(OH)_2$).

A very rigorous Dirichlet boundary condition (zero concentration) induces, at the initial times of the simulation, high flux rates in contact with the nearest dissolving voxels. Therefore, for this case, care should be taken to check the validity of the assumption that the diffusion rate is the rate-controlling mechanism. If the diffusion rate is higher than the dissolution rate, concentration of the dissolving voxel goes below the saturation value. The dissolution rate of $Ca(OH)_2$ depends on the temperature, pH and reactive-surface area.

Development of computational tools for simulating the hydration of cement-based systems, coupled with microstructural development, as well as dissolution due to leaching and further up-scaling reactive-transport modeling, will finally contribute to a better understanding of the ageing phenomenology and thus improve the predicting reliability of the coupled models for concrete-durability assessment.

Acknowledgments This study is part of SUPERCONCRETE Project (H2020-MSCA-RISE-2014, n.645704). Authors like to acknowledge the financial contribution of the European Union as part of the H2020 Programme, and the Marie Curie Actions EU grant FP7-PEOPLE-2010-IEF-272653-DICEM.

References

1. Mainguy M, Coussy O (2000) Propagation fronts during calcium leaching and chloride penetration. J Eng Mech 126:250–257
2. Mainguy M, Tognazzi C, Torrenti JM, Adenot F (2000) Modelling of leaching in pure cement paste and mortar. Cem Concr Res 30-83-90
3. Ulm FJ, Torrenti JM, Adenot F (1999) Chemoporoplasticity of calcium leaching in concrete. J Eng Mech 125:1200–1211
4. Adenot F, Buil M (1992) Modeling of the corrosion of cement paste by deionized water. Cem Concr Res 22:489–496
5. Berner UR (1998) Modeling the incongruent dissolution of hydrated cement materials. Radiochim Acta 44(45):387–393
6. Buil M, Revertegat E, Oliver J (1992) A model of the attack of pure water or under saturated lime solutions on cement. ASTM STP 1123:227–241
7. Ukrainczyk N, Koenders EAB (2014) Representative elementary volumes for 3D modeling of mass transport in cementitious materials. Model Simul Mater Sci Eng 22:035001
8. Atkins PW (1994) Physical chemistry, 5th edn. W. H. Freeman, New York
9. Koenders EAB (1997) Simulation of volume changes in hardened cement-based materials. PhD dissertation, Delft University of Technology, Delft, The Netherlands
10. Wang J, Keener TC, Li G, Khang SJ (1998) The dissolution rate of $Ca(OH)_2$ in aqueous solutions. Chem Eng Commun 169:167–184
11. Rodriguez-Navarro C, Ruiz-Agudo E, Ortega-Huertas M, Hansen E (2005) Nanostructure and irreversible colloidal behavior of $Ca(OH)_2$: implications in cultural heritage conservation. Langmuir 2005:10948–10957

Introduction to Structural Ageing-Specific Functions for Computational Models Based on Synaptic Networks

G. Mavrikas, V. Spitas and C. Spitas

Abstract Individual computational models for various forms of ageing exist, but are neither sufficiently versatile to be directly implementable into complex systems design, nor adaptable to different scenarios of ageing. To address this problem, this work proposes a method to build up models of complex systems, also incorporating ageing, based on a limited number of nominal-operation-related and ageing-specific function primitives, which account collectively for the entire system behaviour. By implementing these functions in synaptic networks, the model enables flexible changes, enrichment, computational effort savings and design-optimization potential. A case study is presented, showing how this method is applied to a helicopter gearbox.

1 Introduction

Due to the nature of all materials tending to return to a thermodynamic equilibrium with the passing of time and the additional influence of the natural and human environment, structures tend to age with repercussions ranging from capital loss due to decreased performance, maintenance, failure and replacement to loss of human lives in the case of some catastrophic failures. Therefore, engineers undertake the task to interpret the behaviour of nature into complex empirical or analytical computational models, in order to use them for inspection and prediction of the lifetime and the performance of structures. The challenging task in terms of design is to manage the complexity of the ageing models in a versatile way; while competent models for many fundamental processes for ageing are known, the design models at hand mostly depend on structure-specific results drawn from specific applications. The result is that, when modifying the design or operating conditions

G. Mavrikas (✉) · V. Spitas
National Technical University of Athens, Athens, Greece
e-mail: mavrikasg@gmail.com

C. Spitas
Technische Universiteit Delft, Delft, The Netherlands

of a structure, the models that describe its ageing behavior do not automatically update/adapt, even though the ageing mechanisms may change—thus, in principle, a new study is required with each design iteration involving major changes (i.e. mode of operation, added parts, different hierarchy), making design for ageing a cumbersome process at best. This is a typical problem in current multi-disciplinary modelling.

The ageing mechanisms, from simple mass loss (e.g. by friction, erosion, corrosion etc.) to material and shape changes (e.g. phase transition, mechanical properties change, cracking and distortion), have been approached through multi-variable governing equations, which may be generalised or case-specific; in the latter case, these must be abstracted before they can be suitable for general use. Often, models are available only at the system level (this is especially true for empirical models developed for specific case studies) and cannot be safely generalised to other systems, subsystems or operating conditions and scenarios. Even when suitably abstract/modular models are available, there is presently no well-defined framework for synthesising them to produce a system-level model, i.e. in the context of designing a new system, so that each time this modelling must be done ad hoc. The resulting lack of versatility and adaptability of the existing modelling techniques pertinent to ageing requires the commitment of substantial time and resources. The theory presented herein aims towards bridging this gap, providing a systematic process for problem identification, system identification and system-level model synthesis for designing for ageing. The vehicle for presenting the theory is a case study of a helicopter gearbox subject to complex operating conditions and a range of ageing phenomena.

Initially, known mathematical models for selected ageing-specific functions for friction-induced wear and crack propagation are presented, alongside the relevant mathematical models describing the nominal operation of important system components (hence not considering ageing). This forms the analytical toolbox for the synthesis of the overall system model.

After this introduction, the abstracted Synaptic Network (SN) primitives [9–11, 16] of the nominal and ageing-specific functions are built. This is supported by an Object-Context-Goal (OCG) classification [12], which serves to identify explicitly the considered operating scenarios and conditions and to streamline and keep the design options and process transparent. On the basis of the system identification and the built SN primitives, the overall system SN is synthesised and presented in graphical form. The SN graph is then parsed into mathematical form by implementing the relevant mathematical models for the nominal and ageing-specific functions. Alternative implementations of the system on the basis of the SN are discussed and the versatility of the presented method is shown.

The introduced design algebra involving SNs and OCG Analysis serves to aid (visual) problem understanding, as well as to provide the advantage of interaction of the design engineer with the design problem at any time, in order to address various design goals and routes (e.g. different case studies, design outcomes, problem-specific constraints, detailed information), enabling a more effective design-space exploration and design optimization.

2   Elements of the Design Algebra

Motivated by the need for a new way of ideas representation balanced between a casual manner and the allowance for validations and implementation in computerized environments, the major constituents of the introduced design algebra are the following:

- *Ideas*: Anything presently or potentially subject to consciousness.
- *Synapses*: Ideas acting as idea links operators.

2.1   Synaptic Networks (SN)

By expressing for any chosen case the pertinent ideas (typically physical manifestations of components and functional interfaces, e.g. contact interfaces between components), they may be implemented as an idea algebra [10] by identifying meaningful links/operations (synapses) between these ideas; the result is a larger model in the form of a Synaptic Network (*SN*) with adjustable levels of complexity, depending on the level of detail at which the identification process is conducted. SNs lend themselves to graphical representation, as shown in Fig. 1. Insofar as each idea, synapsis or subnetwork in a SN can be described by or related to one or more mathematical models, a SN graph can be parsed to create a set of linked models that together form the overall system model.

2.2   Object-Context-Goal (OCG) Analysis

In Object-Context-Goal (OCG) analysis, ideas can be understood and classified as either objects or context or goals. Contrary to goals, which are independently and externally established, objects and context come from characterising the system

Fig. 1 Graphical representation of a SN for a multibody mechanical system

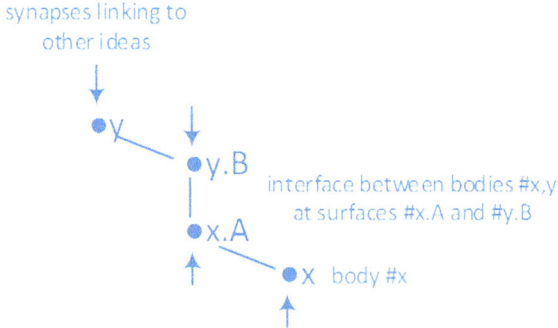

being studied and differ only in the sense that objects can be modified by design intervention, whereas context cannot. Ideas are classified under the Object or Context class during system identification and are by nature coupled by means of the synapses in a SN.

In pursuit of given goals, it is possible to focus on studying particular system behaviours (either desired or not) by considering various scenarios; scenarios basically impose specific object and context states. Scenarios are ideal for studying ageing, in particular, by accelerating ageing processes (intervention to the context) or jumping to semi-aged or fully-aged system states (intervention to the object); otherwise it is very resource-intensive and inefficient to recreate ageing phenomena by simply simulating the full ageing course of an originally 'young' system.

3 System Identification: Case Study of Helicopter Gearbox

For the purpose of the present study, the case of a helicopter main gearbox is discussed.

The main function of a helicopter main-rotor gearbox is to transmit motion (and power) from the fast-rotating horizontal turbomachine engines (rotating at several thousand RPM) to the slow-rotating vertical main rotor (rotating at a few hundred RPM). Another important function is to split the output power between the main rotor (lifting and moving) and the tail rotor (stabilizing). Therefore, the main functions of the gearbox may be described by the following set:

- Merging of the motion/power from two engines (inputs) and splitting them into two main unequal outputs (main rotor and tail rotor).
- Changing the direction of motion from almost horizontal (engines) to almost vertical (main rotor) and horizontal (tail rotor).
- Reduction of the rotating speed from several thousand RPM to a few hundred RPM.
- Increase of the torque from the engines to the rotor.

At the same time, the helicopter gearbox should also ensure:

- Fault-free operation (which involves both operational characteristics i.e. efficiency and structural ones and component structural integrity).
- Acceptable undesired side effects (vibration, noise, temperature rise).
- Controllability either during normal operation or failure of components/subassemblies.

The main system components are identified in Table 1 and graphically shown in Fig. 2.

A quick OCG classification of the gearbox constituents and synapses on a physical level leads to the following rough categorization.

Table 1 Helicopter gearbox components

	Elements	Bearings interfaces		Tooth contacts		
Input shaft assembly	6.1/2	j	k	C		
Second shaft assembly	5.1/2	h	i	C	B	
Central shaft assembly	2	c	d	B	A	D
Planet shaft assembly	3.1/2/3	e	e	D	E	
Main rotor shaft assembly	4	f	g	E		
Tail rotor shaft assembly	1	a	b	A		
Gearbox housing	0					

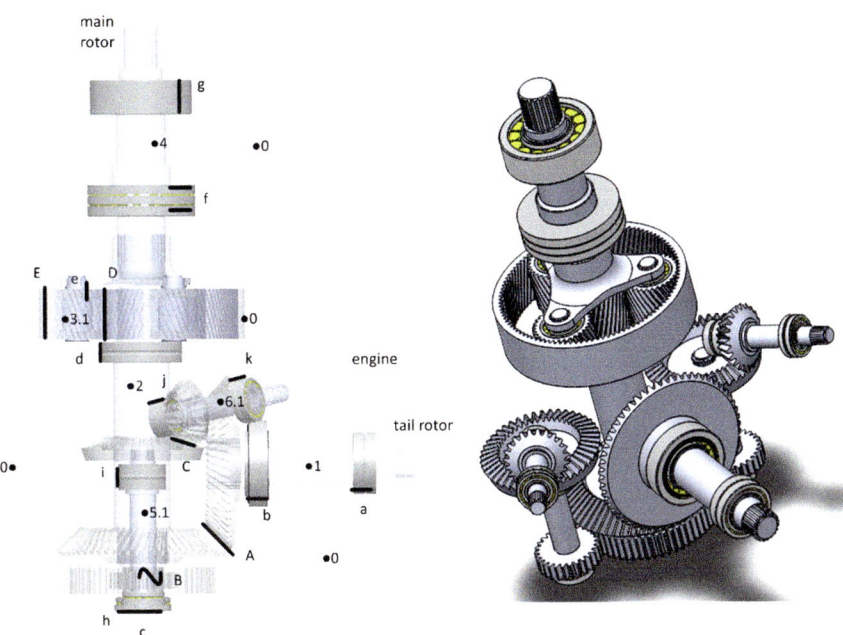

Fig. 2 System identification of a helicopter main gearbox

Object (subject to design/select):

- Gear geometry
- Gear materials
- Shaft/bearing and ancillary component selection
- Mounting methods and internal arrangements
- Lubricants
- Casing

Context (not subject to selection, imposed by the system specifications and natural laws):

- Input/output torque, speed, power
- Direction of input/output shafts
- Performance envelope of the gearbox (including wear resistance and ageing performance)

Since, according to the SN theory Object and Context items are interconnected through synapses, it is important to identify the equations of state/equilibrium and compatibility that connect the above described ideas and link them to ageing-specific metrics (or functions).

Later, kinematics, dynamics and phenomena of ageing will be modelled both mathematically and graphically (SN graphic representation), preparing the ground for the complete SN model.

4 Nominal Operation Models and SN Primitives Implementation

4.1 Gear-Contact Analysis (Geometry, Kinematics)

Since the focus of this work is to model ageing phenomena, it is essential to define the primitive and explicit mathematical model for calculating the dynamic (i.e. forces, torques) and kinematic (i.e. speed, position) characteristics of the cylindrical (i.e. spur/helical) gear transmission. In real applications, like the helicopter gearbox, gear meshing is rarely truly conjugate, thus a state-of-the-art model that addresses gear sets with transmission errors is required.

Spitas and Spitas [14] propose a non-conjugate meshing model, based on the generalised equations for profile tangency. They end up with an explicit scalar equation. This model reduces convergence and in general computational problems, leading to the same or better results from the rest of the state-of-the-art models.

The basic equations that, in conjugate gears, produce the 'law of gearing', are the requirement for geared-power transmission defining a common contact point (Eq. (1)) where profile tangency is observed (Eq. (2)). Even when slipping of tooth profiles occurs in real working conditions, those are still the prime requirements. The slipping of the profiles is modelled through a slip angle (θ_S), expressed in Eq. (3).

$$\vec{r_1} - \vec{r_2} = \vec{a_{12}} \tag{1}$$

$$\left(\frac{\partial \vec{r_1}}{\partial r_1} \times \frac{\partial \vec{r_2}}{\partial r_2} \right) \cdot \vec{x_3} = 0 \tag{2}$$

$$\theta_2 - \theta_{2n} = \theta_s \tag{3}$$

When the vector functions of the two profiles, the rotational matrix around the perpendicular x_3 axis and the center distance vector are added into Eqs. (1–2), Eqs. (4–5) are derived, correlating in this way the gears' individual positions, with their shapes. One more equation (Eq. (6)), derived from a vector analysis of gear 2, completes the constitutive set of the meshing equations.

$$\overrightarrow{R_1}\,\overrightarrow{f_1} - \overrightarrow{R_2}\,\overrightarrow{f_2} = \overrightarrow{a_{12}} \tag{4}$$

$$\left(\overrightarrow{R_1}\frac{d\overrightarrow{r_1}}{dr_1} \times \overrightarrow{R_2}\frac{d\overrightarrow{r_2}}{dr_2}\right) \cdot \overrightarrow{x_3} = U_3(\theta_1, r_1, r_2, \theta_2) = 0 \tag{5}$$

$$\theta_2 = \left[\frac{\overrightarrow{f_2} \times \left(\overrightarrow{R_1}\,\overrightarrow{f_1} - \overrightarrow{a_{12}}\right)}{\left\|\overrightarrow{f_2} \times \left(\overrightarrow{R_1}\,\overrightarrow{f_1} - \overrightarrow{a_{12}}\right)\right\|}\right] \cdot \overrightarrow{x_3} \cos^{-1}\left[\frac{1}{r_2^2}\overrightarrow{f_2} \cdot \left(\overrightarrow{R_1}\,\overrightarrow{f_1} - \overrightarrow{a_{12}}\right)\right] =$$
$$= U_2(\theta_1, r_1, r_2) \tag{6}$$

When combined, the aforementioned equations create the final scalar equation. For a given value of θ_1 (which is the only independent parameter) and the gears' geometry, one can computationally calculate r_1, r_2, θ_2 and θ_S from the model of Eqs. (7–9):

$$U_3(\theta_1, r_1, U_1(\theta_1, r_1), U_2(\theta_1, r_1, U_1(\theta_1, r_1))) = 0 \tag{7}$$

$$r_2 = \left\|\overrightarrow{R_1}\,\overrightarrow{f_1} - \overrightarrow{a_{12}}\right\| = U_1(\theta_1, r_1) \tag{8}$$

$$\theta_2 = U_2(\theta_1, r_1, r_2) \tag{9}$$

where, $\theta_{1,2}$ are the position angles of the gears 1 and 2, θ_S is the slip angle and $r_{1,2}$ are the radii of the point of contact from centers of rotation O_1, O_2 respectively, as per Fig. 3.

The SN implementation of a gear contact is fairly straightforward, as it is simply a contact between one or more pairs of geometrically defined tooth surfaces. If the tooth surfaces of gear 1 are denoted by 1.A and the corresponding surfaces of gear 2 by 2.A, then the SN graph primitive of this interface is as per Fig. 4.

4.2 Dynamics

Keeping the parameters presented in the kinematics analysis and with the need of a dynamic simulation of the indexing errors in spur-gears systems, Spitas and

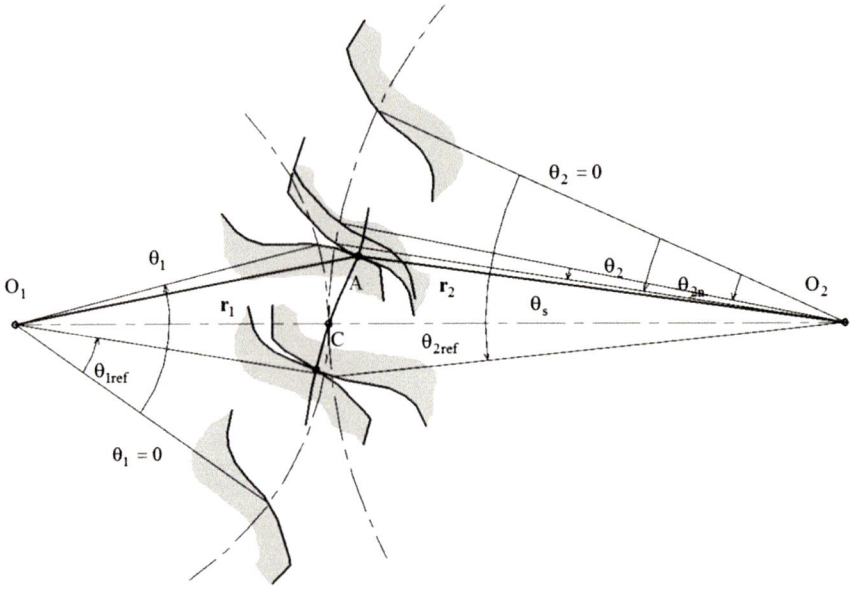

Fig. 3 Gear-contact geometry definitions for the derivation of the meshing equations [14]

Fig. 4 SN graph primitive of
a gear-contact interface

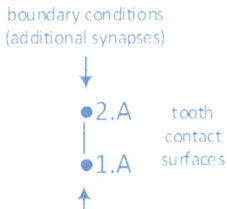

Spitas [13] modelled the effect that the torsional-lateral vibration coupling (F_{hyst}), the tooth friction (F_{frict}) and the tooth separation (F_{elast}) have on the power transmission.

The dynamic model used was adapted to a single-stage spur-gears set, with a rotor and a brake, taking into account the main aforementioned external load for indexing errors. The complication of the model lies in three independent DOFs, i.e. the rotation of the pinion (DOF 1), the gear (DOF 2) and the brake element (DOF 3). In the helicopter-gearbox case study, the modeling of the first two DOFs will be used. As seen in Eqs. (10–12), the external loads are related to the interference (δ_k), the single-tooth bending stiffness (k_k), the damping coefficient for hysteresis (c_{hyst}) and the profile of the gears (f_k).

$$F_{k,elast} = -|\delta_k|k_k\overrightarrow{n_k} \tag{10}$$

Fig. 5 SN graph primitive of
a gear pair

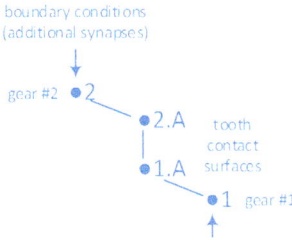

$$F_{k, hyst} = C_{hyst} \frac{d|\delta_k|}{dt} \overrightarrow{n_k} \tag{11}$$

$$F_{k, frict} = |\delta_k| \overrightarrow{f_k} k_k \overrightarrow{m_k} \tag{12}$$

The final set of equations that constitute the non-linear dynamic model alone, is a system of five second-order differential equations which, in conjunction with the tooth-contact analysis, provide the results for the dynamical gear behaviour. This model is presented through Eqs. (13–14), and it is evident that the parameters describing the intrinsic gear behaviour (stiffness, damping, mass etc.) are correlated with the excitation of the system (due to external loads and geometry deviations). For the helicopter-gearbox analysis, the first two DOF are essential.

$$J_j \frac{d^2\theta_j}{dt^2} + D_j \frac{d\theta_j}{dt} - \overrightarrow{x_3} \sum_k \left[\overrightarrow{r_j} \left(F_{elast} + F_{hyst} + F_{frict} \right) \right]$$
$$- E_{shaft} \left(\frac{d\theta_{j-1}}{dt} - \frac{d\theta_j}{dt} \right) - G_{shaft} \left(\theta_{j-1} - \theta_j \right) = 0 \tag{13}$$

$$M_j \frac{d^2 s_1}{dt^2} + C_j \frac{ds_j}{dt} + K_j s_j - \sum_k \left(F_{elast} + F_{hyst} + F_{frict} \right) = 0 \tag{14}$$

The SN implementation of a dynamical gear-pair system is basically the representation of two solid bodies in contact. If the tooth surfaces of gear #1 are denoted by #1.A and the corresponding surfaces of gear #2 by #2.A, then the SN graph primitive of the gear pair is as per Fig. 5.

5 Ageing-Specific Models, Nominal-Ageing Model Coupling and SN Primitives Implementation

Lubricant is the crucial component when considering the most common failure modes in accurately assembled gearboxes. The main functionality of lubricant in gearboxes is the prevention of wear. When referring to ageing in gears, there are two different fundamental mechanisms: the material-removal and the crack-associated ageing.

Scoring/scuffing (adhesive wear) and abrasive wear are the main failure modes of material-removal ageing, thoroughly analysed for decades. These modes are characterised and depend on the classic PV law [2] and concern all the teeth of a single gear. High-shear stress of the lubrication film causes higher temperatures on the contacting tooth flanks, which in turn causes scuffing due to combined high loads, edge contact, excessive vibration and low oil viscosity. This mechanism does not require debris or other foreign particles to exist in the lubricant, however, abrasive wear may manifest itself in the case of heavily contaminated lubricant. By virtue of the zero sliding velocity in the vicinity of the pitch circle and the low Hetzian stress on the tip of the tooth, particularly in light loaded gears, scoring/scuffing mainly appears in between. At the ends of the path of contact, where relative velocity reaches maximum values, this phenomenon has a severe effect.

Regarding the other fundamental ageing mechanism, which is crack-associated modes of failure, pitting and tooth-root breakage are the most common ones. Bending stresses cause breakage of a single tooth, while surface stresses cause pitting. Pitting is generally the most common failure mode in gears, with a mechanism of combined failure modes I and II. Focusing on the pitting mechanism, the sub-surface shear stresses reach the maximum value right below the point of contact, so a mode II crack (in-plane shear) initiates and propagates towards the surface, following the steepest gradient of the stress vectors. At the surface, there is high lubricant pressure at the point of contact (Hetzian stresses at the pitch circle), forcing the lubricant to enter the crack and widen it in mode I (tension). This mixed-mode crack propagates until it reaches the surface again in another adjacent point, having cut a small piece of material (spalling) and leaving a pit on the gear flank. In contrast to the micro-cracking mechanism of pitting, tooth breakage initiates from a stochastically single-tooth root crack in one gear, due to bending stress. Lubricant then enters the crack, assisting the mode I crack propagation from the continuous fatigue bending, leading gradually to tooth deterioration and breakage.

5.1 Material-Removal Ageing (Wear)

Engineers usually cope with wear-failure modes, through the widely acceptable PV relationship, originating from Archard's equation [2], as a computational tool for quantitative results. Almen [1] proposed a variant to the classic PV Law for scuffing resistance, where, apart from the normal pressure and the relative velocity, the distance between the pitch and the point of action, along the path of contact, is inserted, therefore proposing the following PVT Law:

$$(PVT)_G = \frac{\pi n_P}{360}\left(1 + \frac{N_P}{N_G}\right)(\rho_G - R\sin\varphi_t)^2 P_G \tag{15}$$

where N_P, N_G are the number of teeth of the pinion and the gear, n_P is the pinion rotating speed, ρ_G is the radius of curvature at the gear tooth tip, R is the pitch circle radius of the gear, φ_t is the transverse pressure angle and P_G is the maximum Hertzian contact pressure at the tip.

Apart from this failure criterion, researchers who approached these material-removal phenomena, have concluded on the threshold values of lubricant temperature, as well as, on some indexes for scoring/scuffing resistance. First of all, Blok [3] proposed the flash temperature (T_f) (Eq. (16)), containing gears' velocities (V_1, V_2), load magnitude (w), thermal conductivity (λ), heat of surface material (c), material density (ρ) and lubricant-film thickness (a) as the crucial factors. Flash temperature sets the limit of temperature, above which scuffing will probably appear. Later, Kelley [6] and Dudley [4] improved the flash-temperature equation by taking into consideration the surface roughness and the action position on the tooth flank. AGMA Eq. (17) for contact temperature (T_C), which is used for the estimation of the scuffing resistance, contains the scoring index (Eq. (18)), which in this case plays the role of the threshold for scuffing prediction as per AGMA Eq. 17 (1965); and the scoring geometry factor (Eq. (19)).

$$T_f = \frac{1.11\mu w(V_1 - V_2)}{\left(\sqrt{V_1} + \sqrt{V_2}\right)\beta \cdot 2a} \tag{16}$$

$$T_C = T_i + SI \cdot Z_t \cdot \frac{50}{50 - S} \tag{17}$$

$$SI = \left(\frac{W_{te}}{F_e}\right)^{3/4} \frac{n_P^{1/2}}{P_d^{1/4}} \tag{18}$$

$$Z_t = 0.0175 \frac{\left(\sqrt{\rho_P} - \sqrt{\frac{N_P}{N_G}\rho_G}\right)P_d^{1/4}}{(\cos\varphi_t)^{3/4}\left(\frac{\rho_P\rho_G}{\rho_P + \rho_G}\right)^{1/4}} \tag{19}$$

As aforementioned, scuffing occurs at the ends of the path of contact, since at those positions relative velocity (V) and distance (T) reach their maximum values and give rise to the maximum PVT value. In Fig. 6 [9], [15], one can see the isochromatic fringe patterns, which reveal the stress field at the ends of the path of contact in a double-tooth contact condition. With the use of numerical (i.e. FEA) and experimental methods (i.e. photoelasticity, caustics, ultrasounds), the distribution of the stresses along the working involute is revealed.

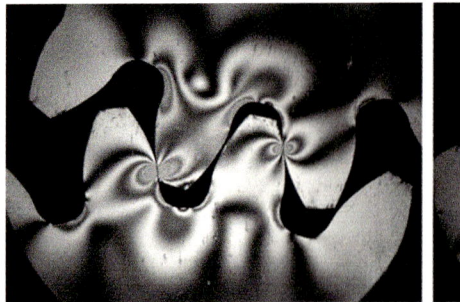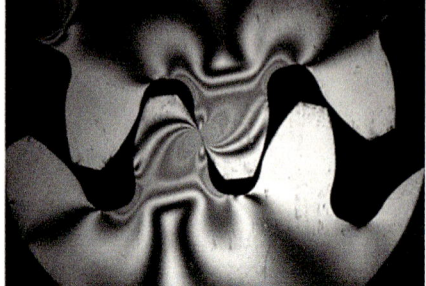

Fig. 6 Photoelasticity study revealing isochromatic stress patterns for double tooth contact (left) and single tooth contact (right) [9]

Fig. 7 SN graph primitive of a surface pair subject to material removal-induced wear

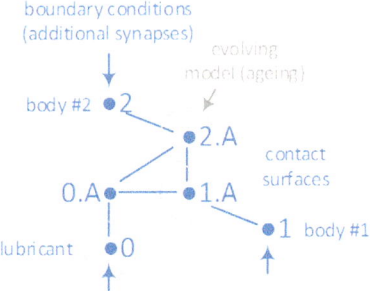

The SN implementation of a surface pair subject to friction-induced wear is basically the representation of two solid bodies in contact, in consideration of a lubricating medium. If the lubricant is represented as body #0, the contact surface of body #1 is denoted by #1.A and the corresponding surface of body #2 by #2.A, then the SN graph primitive of the system is as per Fig. 7. The key difference between an ageing-specific model, such as this, and a nominal operation model, such as shown in the previous section, is that the ageing-specific model affects a self-induced change in the model parameters, thereby evolving itself with time. In the case of wear, this evolution can be beneficial (running-in, polishing), or detrimental (material loss, lubricant contamination, abrasion etc.).

The wear model would be of no value if the anticipated wear would not affect in turn the system operation (tooth-contact patterns and resulting system dynamics), hence, if the system models for wear and nominal operation would not be coupled.

In this case, the coupling is affected by constantly updating the surface geometry definition with the predicted wear figures, to account for the modification/ deterioration resulting from wear (material removal).

5.2 Crack-Associated Ageing (Fatigue)

Pitting is the most detrimental among other failure modes in gears, because, as pits appear on the gear surface, the working tooth surface decreases, having as a result the increase of the Hertzian stresses along the tooth. Rapid deterioration is inevitable and mostly occurs at the lowest point of single tooth contact of the pinion in spur gears. At that position, the critical factor for the pitting phenomenon is the maximum stress combined with the minimum equivalent radius of curvature.

In terms of functionality and gear life, the main difference between pitting and tooth breakage is that pitted teeth could, if required, operate for some time after the manifestation of pitting, contrary to the root-fractured tooth, which, when manifested, impairs the operation of the gearbox as a whole. Since pitting is related to surface stresses and root cracks to bending stresses, AGMA proposes the standard Eqs. (20) and (21) for the calculation of those stresses. Regardless the mode of ageing, Griffith's criterion [5] in Eq. (23) and Paris equation [8] for the crack length in Eq. (24), are the established criteria for a crack-associated failure.

$$\sigma_t = \frac{F_t}{mbJ} K_0 K_u K_s K_m K_B \leq S_{t,cr} = \frac{S_{at} K_L}{S_F K_T K_R} \tag{20}$$

$$\sigma_C = C_P \sqrt{\frac{F_t}{bd_1 J} K_0 K_u K_s K_m C_F} \leq S_{c,cr} = \frac{S_{at} Z_N C_H}{S_H K_T K_R} \tag{21}$$

$$C_P = \sqrt{\frac{1}{\pi \left(\frac{1 - \nu_P^2}{E_P} + \frac{1 - \nu_G^2}{E_G} \right)}} \tag{22}$$

$$K_I = \sigma \sqrt{\pi a} \leq K_{I,cr} \tag{23}$$

$$\frac{da}{dN} = C \cdot \Delta K^m \tag{24}$$

The nature of the gears transmission provide some crack-arrest mechanisms, which are natural mechanisms for stress relief. Firstly, the helicopter gears, and in general all the gears in high-tech and precise transmission gearboxes, are case-hardened instead of through-hardened. The reason is that the non-hardened cores resists crack propagation from the crack-susceptible case, due to their higher fracture toughness. Secondly, the initiation of a root crack increases the compliance to the specific cracked tooth, resulting in load relief overloading the stiffer adjacent teeth.

The SN implementation of a gear subject to fatigue (pitting, root cracking) is basically the representation of a solid body #1 subject to fatigue loading at its surface #1.A. The SN graph primitive of the system is as per Fig. 8. Being an ageing-specific model, it affects a self-induced change in the model parameters, thereby evolving itself with time. In the case of fatigue, this evolution can only be

Fig. 8 SN graph primitive for fatigue (crack/pit development and propagation)

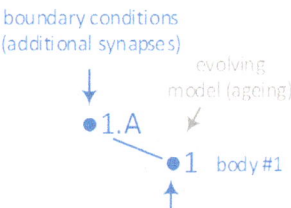

detrimental (loss of strength, loss of stiffness, material loss, lubricant contamination, abrasion etc.).

The fatigue model would be of no value if the anticipated damage (cracks/pits) would not affect in turn the system operation (surface roughness and stiffness changes, resulting lubricant contamination and system dynamics), hence, if the system models for fatigue and nominal operation would not be coupled.

In this case, coupling is achieved by constantly updating the stiffness (and possibly surface-roughness) values for the affected system elements, which in turn affects the dynamic behaviour of the system that leads to further change/ deterioration of the same values.

For helicopter-gearbox design, the most common failure type is pitting or micropitting (frosting), followed by tooth breakage. Although microcracks could appear in tooth roots, they should not be treated as failure signs, but instead they should be monitored as potentially such. Abrasive wear is commonly addressed by lubricant filtration, though new trends propose the use of the high-viscosity engine fuel for lubrication/cooling of the gearbox's mechanical parts, since the power-losses of the gearbox can play the role of pre-heating of the fuel, thereby increasing efficiency.

6 The Complete Synaptic Network

6.1 Joining the SN Primitives and the Sub-networks: SN Synthesis and Audit

Essentially, the complete SN is a cognitive construct that describes our complete (current) understanding of the system and problem being solved, thus including the object and context. As such, it is the union of all the pertinent SN primitives and sub-networks, including those constructed previously in Sects. 4–5. Joining two sub-networks requires that one or more of their ideas overlap, and this is visualized in Fig. 9, where the identical ideas are marked as thin circles of the same color.

Introducing new ideas and joining sub-networks expands the SN. While some of the ideas can be new, in the sense that they can emerge from improvisation/ imagination/inspiration and other creative means, typically most 'building blocks' for the SN are readily retrieved as existing knowledge, i.e. by observing the

Fig. 9 Visualization of joining SNs

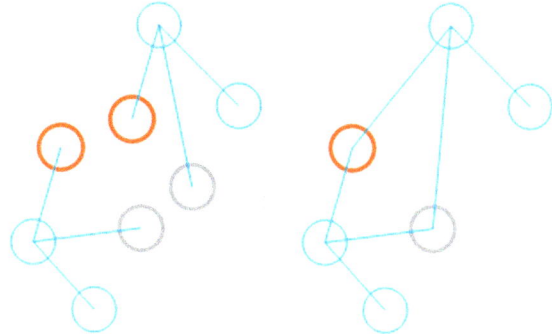

physical links between components and interfaces in similar existing and identified systems (e.g. other gearboxes). This possibility affords a high degree of fail-safety, as each synapsis can be audited to ascertain if it indeed corresponds to a physical link or is otherwise conceptually sound (Fig. 10).

6.2 Parsing the SN: Extraction of the System-Level Mathematical Model

Once the system-level SN is constructed, it can be parsed automatically to identify any matching patterns with predefined SN primitives, like those defined in Sects. 4–5. For each primitive found, the corresponding model is retrieved, adapted to the local parametric definition and appended to the mathematical model of the system. The consistency of definition (naming, links) of the various ideas in the SN serves to couple the various different models. This process is shown visually in Fig. 11. After the parsing is complete, a complete complex mathematical model for the entire system has emerged.

In the present case study of the helicopter gearbox, the system model would comprise tooth-contact models for interfaces A, B, C, D, E; a multi-body dynamics model involving bodies 1, 2, 3.x, 4, 5.x, 6.x, 0; wear models for interfaces A, B, C, D, E; and fatigue models for bodies 1, 2, 3.x, 4, 5.x, 6.x. The possibility to extend the wear and fatigue models to the bearings (interfaces a, b, c, d, e, f, g, h, i, j, k) and to several other subsystems is obvious and is not shown explicitly in the interest of space.

6.3 Reconfiguration of SN—Model Flexibility

The immediacy of the SN-graph representations enables at-a-glance awareness of the main, secondary and auxiliary functionalities of any system configuration:

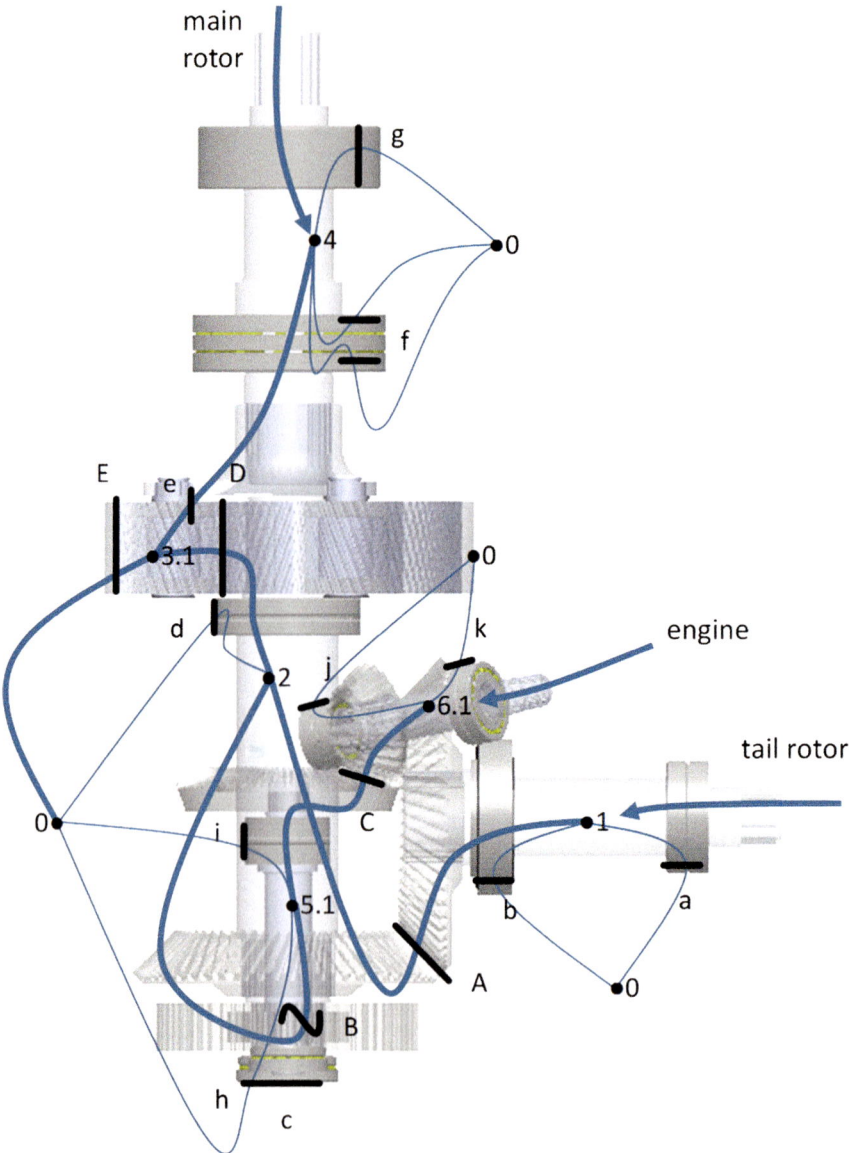

Fig. 10 Synthesized SN of the helicopter's main gearbox system. Thick lines: synaptic paths of main function (motion and power transmission); thin lines: synaptic paths of secondary functions (non-torque load bearing, alignment)

Functions are identified as synaptic paths/loops/subnetworks, e.g. the main function (motion and power transmission to main rotor) is delivered via the synaptic paths 6.x-5.x-2-3.x-4, the secondary function (motion and power transmission to tail

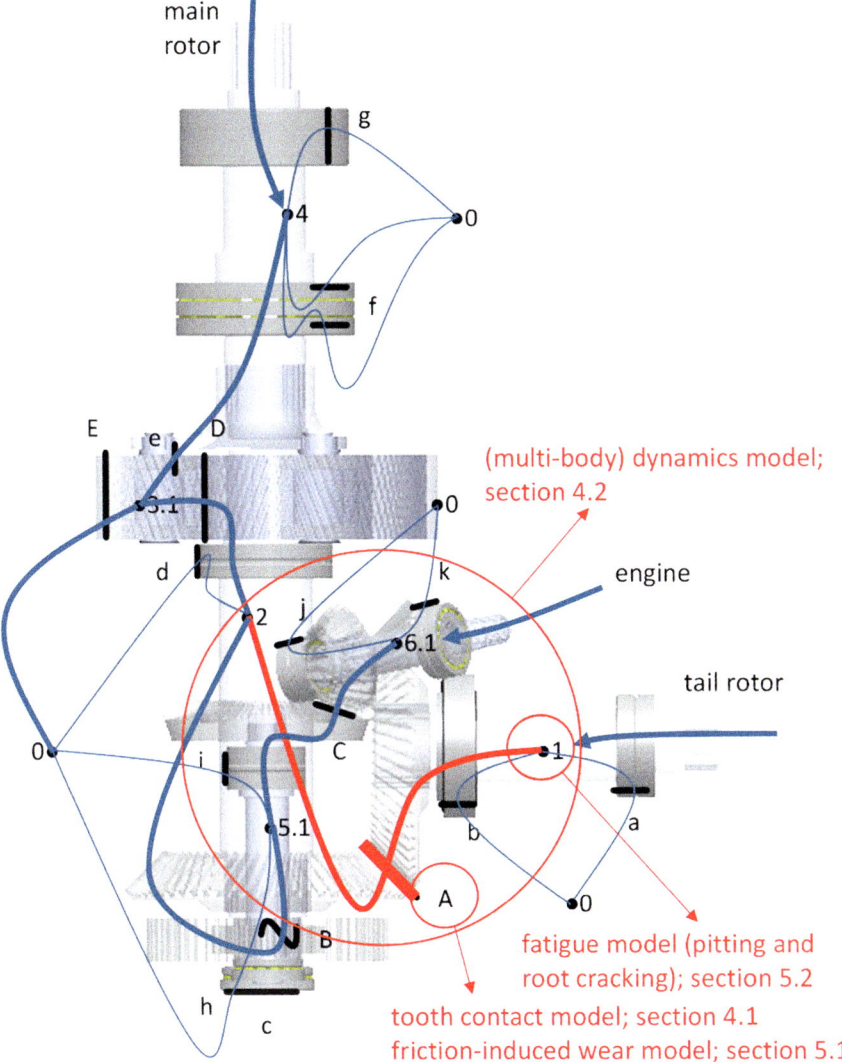

Fig. 11 Visualisation of parsing SNs: automatic extraction of mathematical models from SN primitives (shown here for primitives 1-1.A, 1.A-2.A, 1-1.A-2.A-2). Lubricant subnetwork not shown for readability

rotor) is delivered via the synaptic paths 6.x-5.x-2-1, and an obvious auxiliary function (non-torque load bearing and alignment) is delivered via the synaptic paths 1.b-1-1.a-0 or even the combined 2.c-2-2.d-0-5.x.i-5.x-5.x.h etc. Just as easily, other alternative configurations can be defined by re-arranging the graph, e.g. to change the gear and/or bearing configurations.

For analysis purposes, any 'what-if' scenario can be represented by applying the values of known ideas and computing their effect on the goals or other ideas of interest within the SN. Component positions, orientations, sizes, materials and all other values pertinent to the parametric system definition can thus be changed within the constructed system model. Said flexibility extends to considering any and all parts of the context, such as the operating conditions: i.e. the rotational speed, torque, temperature, lubricant type, cleanliness etc. Both 'young' and 'aged' system states can be simulated. Alternative layouts (e.g. various gear and bearing configurations) can furthermore be tested by reconfiguring and re-parsing the SN, automatically creating a new valid system model each time.

To use the SN for design in, any scenario we may leave a number of values for object/context ideas a priori undefined and calculate them for chosen limit states of the goal criteria. We call these values design Degrees of Freedom (dDOFs). The visual representation of a SN can help recognise the most suitable candidates to serve as DOFs: Ideally such candidates are closely linked to the goals and sufficiently decoupled from each other to not raise conflicts. The number of dDOFs admitting unique solution is in principle limited by the number (and format) of the goal criteria. It is possible that no solution exists for a given set of goals. If a single solution exists for a given problem formulation, then designation of more dDOFs will typically result in an infinity of solutions.

7 Conclusions

Ageing phenomena in structures are complex and necessitate a combination of modelling flexibility to account for different ageing mechanisms and computational rigor. The application of Synaptic Networks (SNs) to the synthesis of computational models based on ageing-specific functional primitives was demonstrated, enabling the assessment of the effect of design solutions on important aspects of ageing behavior (goal). Specifically, through the case study of a helicopter main gearbox, SNs were used to provide:

- Graphical-analytical representations for the modelling realisation of the problem, in terms of the nominal operation and ageing-specific functions that govern different ageing mechanisms
- Emergence of a system-wide model helping in the identification of object-, context- and goal-related ideas, providing a clear overview of what an engineer is able to change/design through his analysis and what are the possible and critical decisions towards this direction.
- Capability of growing the SN by enriching it with more models and deriving more design solutions at any time.
- Capability for systematic reuse of the entire SN or fragments thereof, either by simply changing the values of key parameters, or by evolving the existing model.

Further research is focusing on the implementation of this model in an automated computerized environment [7], providing a universal tool for building, synthesizing and resolving computing SN models of ageing and other engineering systems.

References

1. Almen JO (1950) Surface deterioration of gear teeth. Mech Wear 229
2. Archard JF (1953) Contact and rubbing of flat surfaces. J Appl Phys 24(8):981–988
3. Blok H. Les temperatures de surface dans conditions de graissage sous extremepression. In: Proceedings of the second world petroleum congress, section IV, Paris, vol III, p 471
4. Dudley DW (1954) Practical gear design. McGraw-Hill
5. Griffith AA (1921) The phenomena of rupture and flow in solids. Philos Trans R Soc Lond A221:163–198
6. Kelley BW (1952) A newlook at the scoring phenomena of gears. SAE Trans 61:175
7. Mavrikas G, Spitas V, Spitas C (2015) Functional assembly using synaptic networks: theory and a demonstration case study. In: International conference on engineering design 2015 (ICED15), 27–30 July 2015, Milan
8. Paris PC, Gomez MP, Anderson WE (1961) A rational analytic theory of fatigue. Trend Eng 13:9–14
9. Spitas C (2011) Analysis of systematic engineering design paradigms in industrial practice: scaled experiments. J Eng Des 22(7):447–465
10. Spitas C (2012) Definition of a functional class of ideas for integrated product development and supporting theory. In: 9th international workshop for integrated product development, 5–7 Sept 2012, Magdeburg
11. Spitas C (2013) Beyond frames: a formal human-compatible representation of ideas in design using non-genetic ad-hod and volatile class memberships and corresponding architecture for idea operators. In: International conference on engineering design 2013 (ICED13), 19–22 Aug 2013, Seoul
12. Spitas C (2013) Object-context-goal analysis. In: SIG decision making workshop, Paris
13. Spitas C, Spitas V (2006) Calculation of overloads induced by indexing errors in spur gearboxes using multi-degree-of-freedom dynamic simulation. Proc Inst Mech Eng Part K: J Multi-body Dyn 220
14. Spitas C, Spitas V (2007) Direct analytical solution of a modified form of the meshing equations in two dimensions for non-conjugate gear contact. Appl Math Model 32(1):2162–2171
15. Spitas V, Papadopoulos GA, Spitas C, Costopoulos T (2009) Experimental investigation of load sharing in multiple gear tooth contact using the stress-optical method of caustics. Strain 47:227–233
16. Spitas C, Spitas V, Rajabalinejad M (2013) Case studies in advanced engineering design. In: Proceedings of the 1st international symposium on case studies in advanced engineering design, 17–18 May 2013, Athens

Part VII
Risk Based Models of Ageing

Stochastic Models for Risk and Failure Under Ageing

Jürg Hüsler

Abstract We discuss several stochastic models for the failure of a material or for a risk event. Typically, the failure of a material occurs if the load on the material is higher than its designed maximal load value. However, the material can become weaker by age or use or slow deformation by near-critical loads. Such stochastic models are motivated by realistic situations where the critical load level of a material is not fixed and can change over time. Also, a Bayesian approach will be mentioned with possible applications.

Keywords Risk model · Cumulative shock · Extreme shock · Failure times
Urn model · Ageing model

1 Introduction

A material or a structure is constructed based on a certain design. It is expected that the material will function well as long as it is not overloaded or over-stressed by some impact factors. For bridges it is the load, for wind turbines it is the strong wind or storm, for a steel plate it is the force or load, for a aircraft it is the acceleration in flight programs. The material at the beginning is possibly well constructed and in good shape that is certainly checked before use. Unfortunately, with time and use the construction gets weaker. The steel or concrete of a bridge or tower will be weakened by the weather impact; the military aircraft is less robust after several training flights with many large accelerations depending on the flight programs. It means that, with time, the material cannot be loaded or stressed as when new.

Each material has a designed lifetime during which no failure should occur. For example, an aircraft has a given number of flight hours that denotes its (technical) lifetime. Since failure can always happen, one should know the probability of such an event during the designed lifetime. The lifetime is not a fixed time but a random

J. Hüsler (✉)

Department of Mathematics and Statistics, University of Bern, Bern, Switzerland
e-mail: huesler@stat.unibe.ch

K. van Breugel et al. (eds.), *The Ageing of Materials and Structures*,
https://doi.org/10.1007/978-3-319-70194-3_16

219

value. Hence, it would be better to know the distribution of the time until failure or the so-called survival distribution.

Typically, one does not have many data at the beginning when designing the material and later when using the material. Often, one has physical models and engineering methods to determine the construction of the material for an intended maximal load or strength. Other models are used in accelerated statistical testing. In some cases, one can collect data on the survival time or time to failure, but the data are related to some conditions only. For other situations, one applies generalization or extensions which might be debatable.

Ageing has an impact on the material and on its maximal load below which no failure will happen. However, it is possible that the large loads have an impact on this crucial load, not only the natural ageing. We think that some loads can change the crucial load value, in particular if a load is close to the crucial one. For instance, a large stress or acceleration may produce a crack in the neighbourhood of a bolt in an aircraft which weakens the strength of the aircraft.

In many cases, the material (an aircraft, a concrete bridge, a steel construction) is controlled in regular time periods to prevent a failure during the inspection periods because of small or moderate damages, use or ageing. Such controls have an impact on the behaviour and the failure probabilities. This should also be included somehow in a stochastic model. We will deal with such probabilistic models for failure events in Sect. 2.

Simple and more advanced risk and failure models occur in many different fields. We mention for instance the insurance business where risk has to be modeled in their applications. A simple risk model consists of the premiums and the claims. The claims occur at random times; the premiums are paid in a more or less regular way. The claim sizes are random with a certain underlying distribution. The premiums should cover the claims at any time. With a certain reserve, the risk process (premiums plus reserve minus claims) should not become negative, which means that the last claim is no longer covered when this happens. This is the so-called (technical) ruin event. There exists several different models for the risk process in the insurance field. Some may be used also as a model for the ageing process.

2   Simple Failure Models

A failure of a structure can happen in various ways. It might be that many small or moderate shocks sum up to a certain critical level when the load becomes too large and the structure breaks down or fails with the next load. It might be also that the structure is not harmed as long as the load is lower than the designed level, but it collapses if the load is larger than the critical value. These two scenarios are called the cumulative shock and the extreme shock model, respectively. The shocks are arriving irregularly over time. These arrival times are best modeled as random variables.

Mathematically, it is described as follows. Let the shock sizes or load values be denoted by $X_i, i \geq 1$, which happen at certain random time points T_i. Let u denote the designed critical value. In the cumulative shock model, a failure occurs the first time at T_k if the partial sum

$$S_k = \sum_{1 \leq i \leq k} X_i > u, \quad \text{but} \quad S_i \leq u \text{ for } i < k.$$

In the extreme shock model, this happens the first time at T_k if

$$X_k > u, \quad \text{but} \quad X_i \leq u \text{ for all } i < k.$$

Then the number of shocks τ until failure is defined as $\tau = \min\{k : X_k > u\}$ in the extreme shock model, or $\tau = \min\{k : S_k > u\}$ in the cumulative shock model. Let us call τ simply the failure time. The lifetime of the system is denoted by T_τ, since no failure occurs before T_τ (Figs. 1 and 2).

One can also combine both failure events into one, i.e. a failure occurs at T_k if either $S_k > u$ or $X_k > u$ happens. Hence, $\tau = \tau(u) = \min\{k : S_k > u \text{ or } X_k > u\}$. There is no need to assume that the two critical levels are equal. We may define more generally $\tau = \tau(u, v) = \min\{k : S_k > u \text{ or } X_k > v\}$. Such models are called mixed shock models. Motivating examples are given in [14]. For instance, a nuclear power plant emits some radioactivity in regular daily operation. After some time, this amount is larger than a fixed threshold u. On the other side, an accident may happen such that a large amount of radioactivity is emitted, which exceeds another threshold v being dangerous or deadly for people in the power plant.

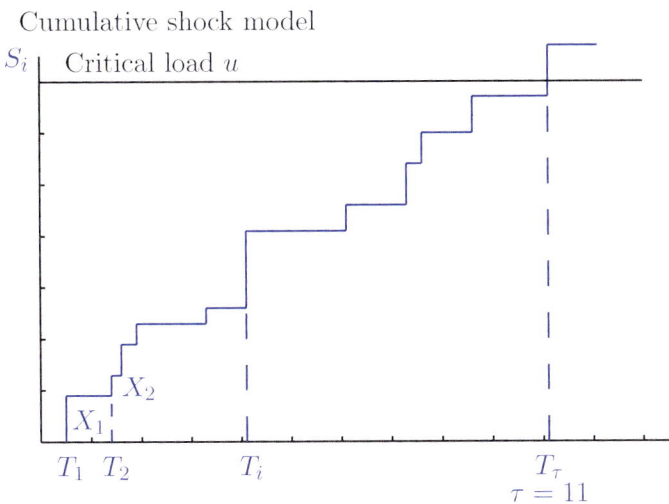

Fig. 1 Cumulative shock model with shocks X_i

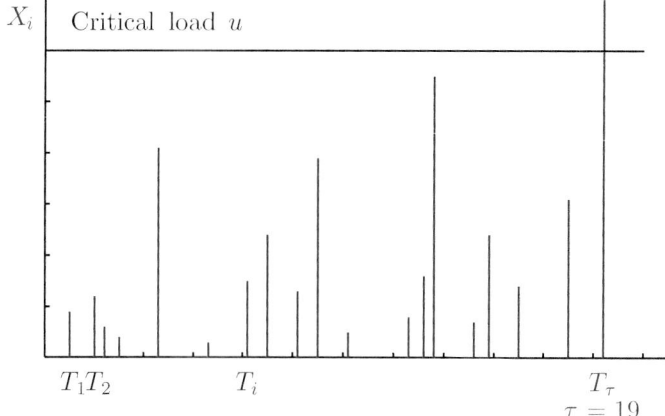

Fig. 2 Extreme shock model with shocks X_i

Depending on u and v, either the extreme shock or the sum shock may dominate the failure. Hence u and v have to satisfy an assumption such that both shock models contribute to the mixed model as $u, v \to \infty$. We assume the upper endpoint of X_i to be infinite, for simplicity. We must have that $p_v = P(X_i > v)$ and $1/u$ are of the same order for $u, v \to \infty$.

For these simple models, one can derive the distribution of the failure time τ and the lifetime T_τ under further assumptions on the random variables X_i and $Y_i = T_i - T_{i-1}$, the inter-arrival time.

We assume that the shocks X_i are independent and identically distributed (iid) with distribution F, and also that the inter-arrival times Y_i are iid. It is not necessary to assume that the shock X_i and the inter-arrival times Y_i are independent. Denote by μ_X, μ_Y and σ_X^2, σ_Y^2 the means and variances of X_i and Y_i. Then, from renewal theory we get:

Result 1: Cumulative shock model

$$\frac{\tau(u)}{u} \to \frac{1}{\mu_X} \quad \text{as } u \to \infty, \text{ almost surely}$$

where the limit equals 0 if $\mu_X = +\infty$.

(i) If μ_X and μ_Y are finite, then:

$$\frac{T_{\tau(u)}}{u} \to \frac{\mu_Y}{\mu_X} \quad \text{as } u \to \infty, \text{ almost surely.}$$

(ii) If, in addition, $\sigma_X^2 < \infty$, $\sigma_Y^2 < \infty$, and $\sigma^2 = \text{Var}(\mu_X Y - \mu_Y X) > 0$, then:

$$\frac{T_{\tau(u)} - \frac{\mu_Y}{\mu_X} u}{\sqrt{\mu_X^{-3}\sigma^2 u}} \to N(0,1) \quad \text{as } u \to \infty, \text{ in distribution.}$$

Suitable references with proofs are [11, 12] and [16].

Result 2: Extreme shock model

Let $p_u = P(X_1 > u) = 1 - F(u) \to 0$ as $u \to x_F$ (endpoint of the distribution F). The distribution of $\tau(u)$ is geometric: $P(\tau(t) > k) = (1 - p(u))^k$ with mean $1/p_u$. Hence:

$$p_u \tau(u) \to \text{Exp}(1) \quad \text{as } u \to x_F, \text{ in distribution.}$$

Since, in this case, $T_{\tau(u)}$ is a stopped random walk also, and if μ_Y is finite, then:

$$p_u T_{\tau(u)} \to \text{Exp}(\mu_Y) \quad \text{as } u \to x_F, \text{ in distribution.}$$

Details, further results and proofs are given in [14] with further references.

Result 3: Mixed shock model

If μ_X and μ_Y are finite and $v = v(u) = F^{-1}(1 - \theta/u)$, the $(1 - \theta/u)$-quantile of F, for some $\theta > 0$, then:

$$\frac{\tau(u, v)}{u} \to Z \quad \text{as } u \to \infty, \text{ in distribution}$$

where the density of Z is

$$f_Z(y) = \theta e^{-\theta y}, \quad \text{for } 0 < y < 1/\mu_X, \text{ with a point mass } P(Z = 1/\mu_X) = e^{-\theta/\mu_X},$$

and
$$\frac{T_{\tau(u,v)}}{u} \to \mu_Y Z \quad \text{as } u \to \infty, \text{ in distribution,}$$

$$\frac{S_{\tau(u,v)}}{u} \to \mu_X Z \quad \text{as } u \to \infty, \text{ in distribution,}$$

$$\frac{X_{\tau(u,v)}}{u} \to 0 \quad \text{and} \quad \frac{\max\limits_{1 \le k \le \tau(u,v)} X_k}{u} \to 0, \quad \text{in probability}$$

Details, further results and proofs are given in [14].

Shock Models under Ageing

For instance, the financial models do work with the cumulative approach. The cumulative shock model can be viewed as a risk model under ageing. The shocks during the lifetime are summed up until the sum is larger than the designed critical value. The life is spent. A small shock can ruin the structure. One observes a failure if $X_k > u - \sum_{j<k} X_j = u(k)$, a decreasing threshold function.

If the shock X_i is not directly harming the material, one may take a function g of X_i which denotes the impact on the structure. But this can be rewritten by putting

$g(X_i)$ instead of X_i, which changes only the distribution of the random variables in the sum S_i. Mathematically, this introduces only a small change.

Such models consider a certain random ageing effect, because the designed critical value u is varying from shock to shock. With age, the level u is decreasing in time where $u(k)$ is random. However, not every shock has an impact on the structure. We may pose that a shock has an impact only if its value is above the level a. Then the threshold decreases somehow, by a certain amount, as Fig. 3 shows it. It shows the threshold $u(k + 1) = u(k) - \gamma 1(a < X_k)$.

In [13] we discussed also shock models where, at the beginning of a structure, the threshold may improve because of run-in. A shock is improving if it is above b but lower than a (say). After the first shock of a certain value (say above a), the structure cannot improve anymore (as in the example of the aircraft where the shock implies the first crack). Such a boundary $u(k)$ is shown in Fig. 4. Here the threshold values $u(k)$ are given by:

$$u(k + 1) = u(k) + \beta 1(b < X_k \leq a, X_i \leq a, i \leq k) - \gamma 1(a < X_k)$$

where β is the change of the critical threshold after an improving shock X_k and γ the change after a harmful shock.

Other decays of $u(k)$ can also be considered where the age is also incorporated. Assuming a certain pattern for this decay, one can again derive the distribution of the failure time τ or life time T_r. So, one would have to adjust the level u at the shock time k, depending on the age T_k and the previous shocks. Such a general function with a linear ageing trend after a run-in time θ could be:

$$u(k) = u - d(T_k - \theta) + \beta \sum_{j<k} 1(b < X_j \leq a, X_i \leq a, i < j) - \sum_{j<k} g(X_i - a)$$

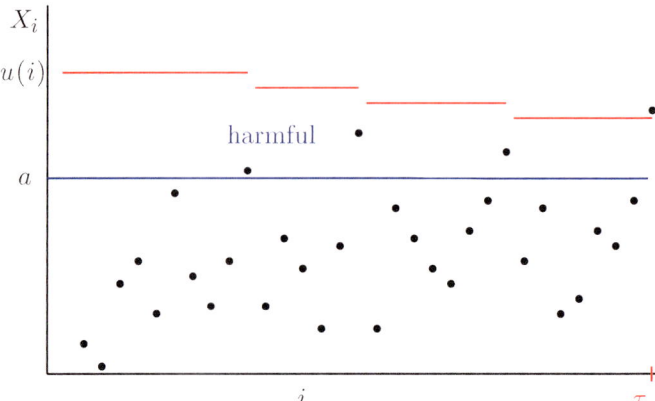

Fig. 3 Example of varying thresholds because of harmful shocks or age

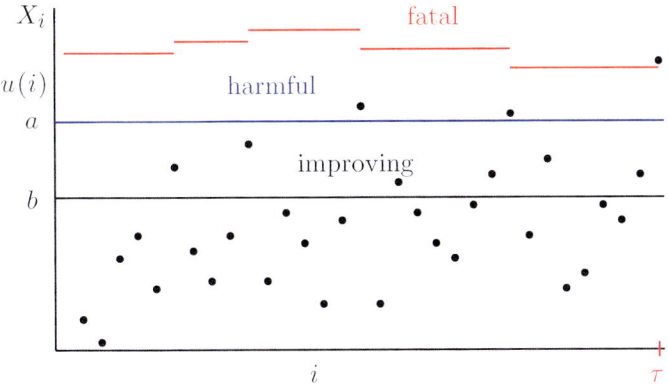

Fig. 4 Example of varying thresholds because of improving and harmful shocks or age

with a (positive) linear or nonlinear function $g(\cdot)$ as impact of the harming shocks with $g(x) = 0$ for $x < 0$. Some theoretical results for such cases are given in [13] for $d = 0$ (no ageing trend). If $d > 0$ then the inter-arrival time plays a crucial role in the threshold, which makes the derivation of results more involved. For example, we present the result for a linear trend with $d > 0 = \theta$ without assuming an improvement or a harming of shocks ($\beta = 0 = g(\cdot)$). Thus, $u(k) = u - d\,T_k$ is a random threshold function. Assume that the interarrival-times are independent of the shocks X_i. Denote the density of the iid. interarrival-times by $g_T(t)$. Then:

$$P(\tau > 1) = \int_0^\infty F_X(u - d \cdot t)g_T(t)dt$$

$$P(\tau > 2) = \int_0^\infty F_X(u - d \cdot t_1)g_T(t_1)\int_0^\infty F_X(u - d(t_1 + t_2))g_t(t_2)dt_2dt_1.$$

The distributions of the failure time or of the lifetime of this model are straightforward to derive numerically by simulations. We assume for our following simulation that X_i and Y_i are iid, exponential random variables, with $\lambda = 1$, since different λ's can be dealt with by transformation and changing the values u and d. We set $u = 10$ and $d = 0.1, 0.2$ and 0.3. We simulated 10,000 values for the failure time τ and derived the simulated survival probability $P(\tau > t)$ (Fig. 5). In the same way, one can also derive with the same simulation the distribution of T_τ (not shown).

However, if we assume that the critical levels $u(k)$ are deterministic and the shocks are independent and identically distributed with distribution F, then it is straightforward to derive the distribution of non-failure until the k-th shock:

$$P\{\tau > k\} = P\{X_i < u(i), i \le k\} = \prod_{1 \le i \le k} F(u(i)).$$

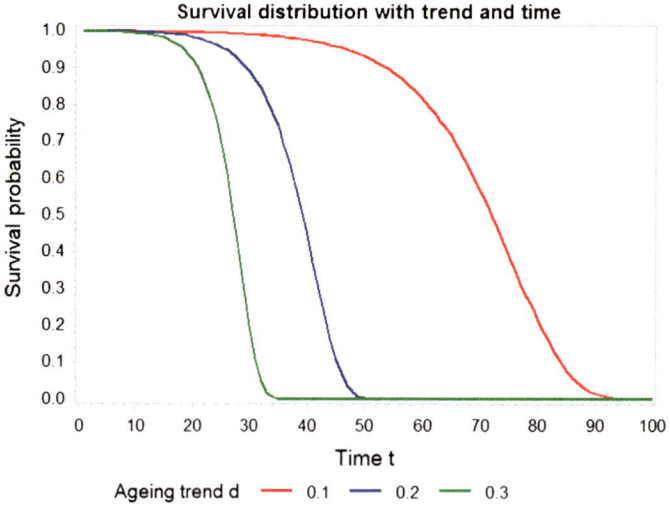

Fig. 5 Survival distribution $P(\tau > t)$ based on 10,000 simulations with $u = 10$ and different trends d where $u(k) = u - d \cdot T_k$

If the levels $u(i)$ are large, one can derive an approximate (asymptotic) distribution for τ and then also for the lifetime T_τ. Assume that $u(i)$ is decreasing only if a shock is non-fatal, but damaging, i.e. $X_i \in (a, u(i))$. In [15] the distribution of τ is derived:

$$P\{\tau > k\} = \sum_{j=0}^{k} \binom{k}{j} F^{k-j}(a) \prod_{i=0}^{j-1} [F(u(i)) - F(a)]$$

where j indicates the number of damaging shocks until m and a denotes the lower boundary of a damaging shock. If the levels $u(i)$ are large, one can derive the asymptotic distribution. See [15], also for some further examples.

If we introduce the inspection with repairing in a regular scheme, for example, always after m shocks (for instance after m flight programs), we would only change the critical level function $u(k)$. We would replace for instance the function: $u(k) = u - \sum_{1 \le i < k} g(X_i - a)$ by the function:

$$u(k) = u - \sum_{j \ge 0} 1(jm < k \le (j+1)m) \sum_{jm \le i < k} g_j(X_i - a).$$

This would assume that after the j-th inspection and necessary repair, the material is again as new.

If an ageing effect should be included, we would add an ageing term in the function $u(\cdot)$. For instance we may define

$$u(k) = u - \sum_{j \geq 0} 1(jm < k \leq (j+1)m) \left[b_j + \sum_{(j-1)m \leq i < k} g_j(X_i - a) \right]$$

where b_j denotes the ageing effect induced after the j-th cycle.

Assuming a simple situation with deterministic decreasing critical levels $u(k)$, an inspection after m shocks, and after repair 'as new', the distribution of τ is

$$P(\tau > k) = \left[\prod_{i=1}^{m} F(u(i)) \right]^j \prod_{i=1}^{k-jm} F(u(i)), \quad \text{for } jm < k \leq (j+1)m.$$

Assuming numerical values for the critical levels and a distribution function F for the shocks, we can easily determine numerically the distribution of τ. Let $u(k) = u - d(k-1)$ and the $F(x) = 1 - \exp(-\lambda x)$. The numerical values of $P(\tau > t)$ are plotted in Fig. 6, where we let $d = 0.2, u = 10, \lambda = 1$ and some inspection parameters m (with $m = 0$ indicating no inspection). The impact of the inspections is nicely visible.

From a probabilistic point of view, these models are simple. From a practical point of view, we would need a good amount of information for modeling and deriving the distributions of τ and T_τ for safety reasons. However, this is true in all cases of deriving the risk of a failure. However, it is straightforward to investigate several scenarios for estimating the risk of a failure with such models.

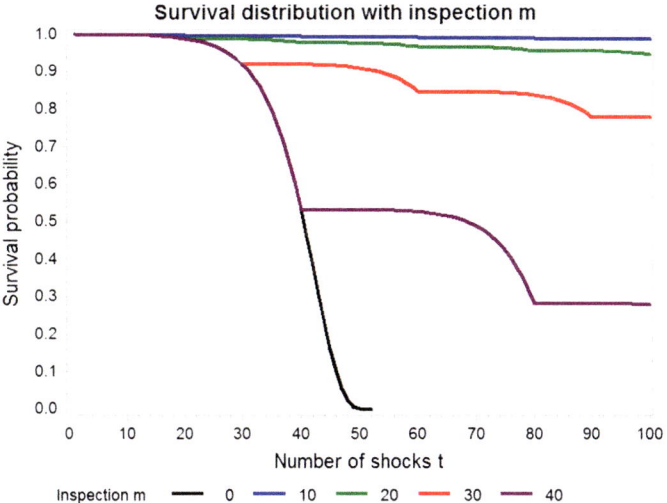

Fig. 6 Survival distribution with inspection intervals m ($m = 0$ no inspection) and deterministic linear trend

3   Urn Models for the Ageing Process, a Nonparametric Approach

Another rather simple but flexible approach for modeling failure and risk with ageing is given by the Pólya urn models. Urn models are basic tools in probability and have been well-known for centuries. Urn models consist of urns with balls of different colors. A ball is randomly drawn from the urn. Depending on the color of the ball, a certain defined event happens. The ball is possibly replaced or even replaced together with additional balls of other colors.

In the simplest form, one has one urn with balls of two colors, say white and black. If a black ball is drawn, one says, the system fails, if a white ball is selected, then the system continues to operate. If the drawn ball is replaced, the content of the urn is not changed, and one has independent drawings with identical probabilities of the choice of a white or a black ball, respectively. This simple model relates to the extreme shock model with constant u where $P(X_i \leq u)$ is given by the percentage of white balls in the urn.

One may have also several urns in parallel, which indicate different components of failure. One can draw a ball from each urn. If we get at least one black ball, we say the system fails. If we draw only white balls, no failure occurs and they are all replaced maybe together with some additional balls of white and black colors. By adding balls, we change the content and the failure probability, which can be related to an ageing effect.

One may also consider the case in which one draws one ball from only one randomly chosen urn. It implies that the probability of a failure is not the same at each drawing, conditionally on the selected urn, if the urns have different numbers of white and black balls. One may also consider dependent urns which means that a drawing from one urn has an impact on the number of balls in other urns, by adding white or black balls in all or some of the urns. This would be appropriate for systems with several components that influence other components.

These models are rather flexible and allow also to model the ageing by adding black balls together with white balls after a white ball is drawn. The ageing effect depends in this case on the number of drawn white balls, not on the time. But the time T_k could be added also.

More flexibility would give us an urn with balls of three colors. Let us say we have the three colors: black, red and white. A black ball indicates again a failure; a red one stands for an ageing step or a non-fatal damage which lowers the critical load value u; and a white one indicates no harm to the structure or material. Using our knowledge of a particular structure, we might start the urn with a prior distribution of the number of balls in the urn. Say, we select w white balls, r red, and b black ones. Then the ageing is modeled by the reinforcement. If one selects a white ball, indicating a non-harmful event, this ball is just replaced. The chance of a harmful or of a failure event is not changed. But if we think that the material may also become stronger or safer, one might model this simply by adding white balls to the urn. If the chances of a failure changes by ageing, then we add some black or red balls after a white has been drawn. If a black ball is selected, then the structure fails. If a red ball is drawn, then the structure has changed indicating also possible ageing, and the chance of a failure is getting larger. The critical level should no more be the same. Hence, we replace the drawn red ball by adding some more red and black ones. This results in a reinforcement matrix R indicating the numbers of balls being added to the urn after replacing the drawn ball. A non-trivial reinforcement allows also to add white balls if a white ball has been selected. This allows that the system gets safer. A balanced reinforcement matrix is for instance

$$R = \begin{pmatrix} \theta & 0 & 0 \\ 0 & \delta & \lambda \\ 0 & 0 & \theta \end{pmatrix}$$

where we let $\lambda = \theta - \delta \geq 0$, $\theta > 0$ and $\delta > 0$. If we let $\lambda = 0$, hence $\delta = \theta$. Thus, a red ball is not replaced together with another black ball. The urn is always reinforced by additional balls of the same color as the drawn one. The urn content is changing depending on the randomly selected balls. The number of white and red balls are increasing before a failure. Hence the structure becomes safer. A red ball has no impact on the failure and is therefore not needed in the urn. Thus one should let $\lambda > 0$, so that the process shows some dependence between the red and black balls. A drawn red ball changes the failure probability because of the added black balls. The probability of a failure (drawing a black ball) is, after the k-th drawing with $j (j \leq k)$ selected red balls:

$$(b + j\lambda)/(w + r + b + k\theta)$$

where w, r, b denote the numbers of white, red, and black balls, as the prior distribution, before drawing any ball. Note that the urn is changing randomly, so a random ageing effect applies here.

Such models can be easily simulated to derive results on the number of drawings until failure, the failure time, or the lifetime. One can use prior knowledge from the design and material properties to select w, r, b and θ, λ. Simulations enable one also to investigate the impact of selecting the prior w, r, b, θ, λ on the failure times. In the following derivation, we assume the balanced reinforcement matrix

$$R = \begin{pmatrix} 3 & 0 & 0 \\ 0 & 2 & 1 \\ 0 & 0 & 3 \end{pmatrix} \tag{1}$$

with initially eight red balls and two black ones, with different secenarios for the number of initial white balls, $w = 990, 1190, 1390$, and 1590. The estimated distribution of the failure time is based on 3,000 simulations resulting in estimated medians: $280.5, 341, 380$, and 447.5 shocks for the four scenarios.

If one has an inspection plan, one can say that, after m drawings, the urn is replaced by the same urn content as at the beginning ('repaired is as new') or by a

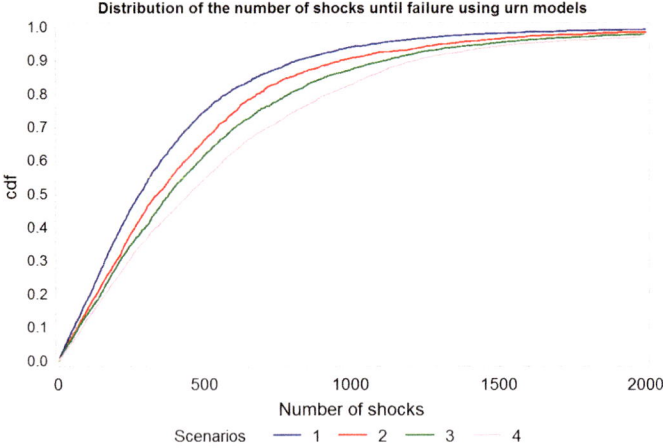

Fig. 7 Distribution of number of shocks until failure using the urn model where the number of initial white balls $w = 990 + (l - 1) * 200$ for the 4 scenarios, and initially $r = 8, b = 2$, and the reinforcement is given in (1) by $ww = 3, rr = 2, rb = 1$ and $bb = 3$

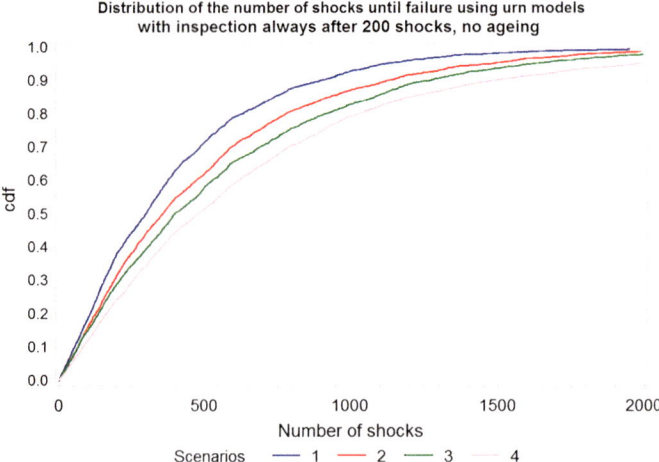

Fig. 8 Distribution of number of shocks until failure, using the same urn model as in Fig. 7 with inspection after 200 shocks

slightly changed content (to model the ageing with inspection and repairing). The estimated medians are now larger: 301, 358.5, 399, and 476 shocks.

We can add easily also an ageing effect, by adding after each inspection and repair some red balls. In the following simulations, we added five red balls after each inspection as an ageing effect. The medians are now smaller: 292, 340, 386.5, and 449.5 shocks (Fig. 9).

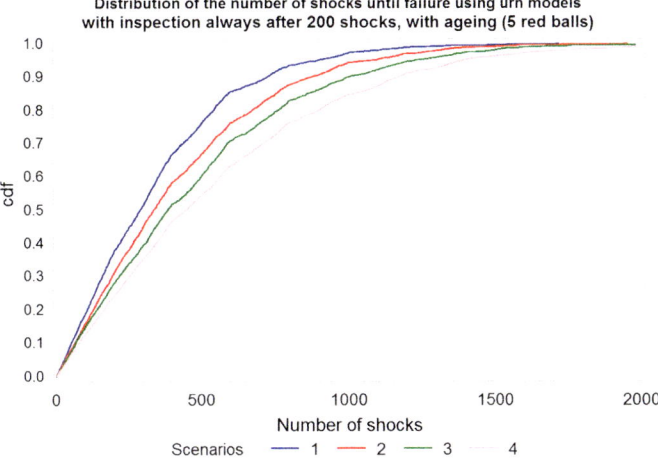

Fig. 9 Distribution of number of shocks until failure using the urn model with inspection after 200 shocks as in Fig. 8 now with ageing (adding five red balls).

Depending on the assumptions of the reinforcement matrix R, [1–8] derived theoretical results on the failure distribution for several special and more general urn models and showed some applications.

4 Modelling with Extreme Value Distributions

Let us consider the following corrosion example for the third approach. This is a statistical approach. The deepest hole in a tube or a metal plate should be measured in an inspection, e.g., in the inspection of the tube for oil transport or pumping from the ground. One samples the tube at several sites following an inspection plan and measures the thickness of the tube material. Based on the inspection data, one should estimate the distribution of the smallest thickness value of the tube at any site. This is one of the many applications of extreme value theory. It says that the distribution of the smallest tube thickness (based on a large sample of observed values of the tube thickness) can be approximated by the generalized extreme value distributions. If one uses all the smallest values lower than a given threshold, one can approximate the tail of this (conditional) distribution by a so-called generalized Pareto distribution which is related to the generalized extreme value distributions.

The class of extreme value distributions G_γ for maxima is well-known (see, e.g. [9, 10, 17–19]):

$$G_\gamma(x) = \exp\left(-(1 + \gamma(\frac{x - \lambda}{\sigma}))^{-1/\gamma}\right), \quad \text{where } 1 + \gamma(\frac{x - \lambda}{\sigma}) > 0$$

with λ and γ the location and shape parameters (real values), and $\sigma > 0$ the scale parameter. For $\gamma = 0$, we have by continuity

$$G_0(x) = \exp(-\exp(-\frac{x-\lambda}{\sigma})).$$

For $\gamma < 0$, the distribution has a finite upper endpoint $\lambda - \sigma/\gamma$; otherwise the upper endpoint is infinite.

For minima, we obtain the class of (minima) extreme value distributions $\tilde{G}_\gamma$:

$$\tilde{G}_\gamma(x) = 1 - G_\gamma(-x)$$

using the relation $\min(X_i, i \le n) = -\max(-X_i, i \le n)$. The parameters λ, σ and γ are estimated using one of the many estimators as moment estimators, maximum likelihood estimators, probability weighted estimators, the Hill estimator and further variants to correct the possible bias (see, e.g., [14, 15]).

These three parameters may change because of age. The ageing process can be modeled into the parameters. After several inspections, one may have sufficient data to model or estimate the ageing process, by assuming that the parameters $\gamma(t), \sigma(t)$, and $\lambda(t)$ depend on the (inspection) time or age t in a certain way, as in the former sections.

If one uses in the data analyses, all data above a threshold, or as in our example of the tube thickness, the data below a threshold, one applies the so-called peaks over threshold approach (POT) where the tail of the distribution can be approximated by a generalized Pareto distribution. This family of distributions W_γ for POT is related to the extreme value distributions by

$$W_\gamma(x) = 1 + \log G_\gamma(x) = 1 - (1 + \gamma\frac{x-\lambda}{\sigma})^{-1/\gamma} \quad \text{if} \quad \log G(x) > -1$$

which depend on the three parameters γ (shape), σ (scale), and λ (location). Again, the three parameters can be modeled depending on time t for ageing.

For our tube thickness example, one has to use the approach for peaks lower threshold (PLT) or the mentioned transformation for minima. The class of distributions for the PLT approach for minima is

$$\tilde{W}_\gamma(x) = (1 - \gamma\frac{x+\lambda}{\sigma})^{-1/\gamma}$$

where $x \le -\lambda$ if $\gamma \ge 0$, or $-\lambda + \sigma/\gamma \le x \le -\lambda$ if $\gamma < 0$. If $\gamma = 0$, we have $\tilde{W}_\gamma(x) = \exp((x+\lambda)/\sigma))$ for $x \le -\lambda$.

For our tube example, we think that the lower boundary is naturally 0. Hence, the shape parameter γ should be negative (or 0). Since the boundary of the generalized Pareto distribution for negative γ is depending on γ, we may use another parametrization of this class of distribution which is simply a Beta distribution for $\gamma < 0$. We have that the thickness data below a threshold can be modeled with the

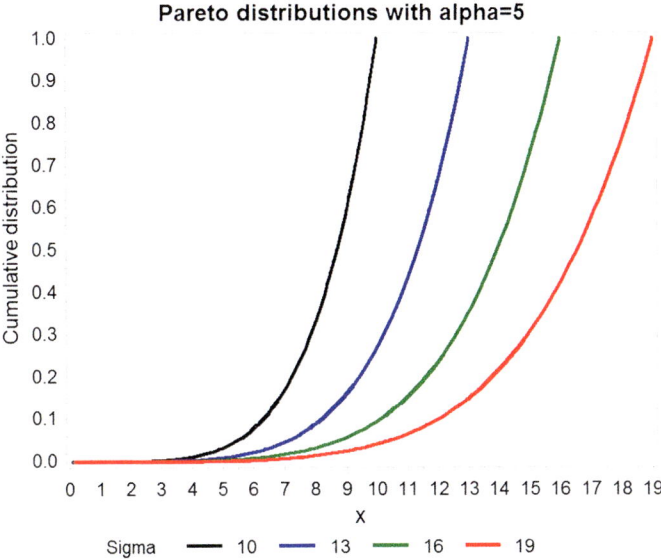

Fig. 10 Pareto distributions with $\alpha = 5$ and $\sigma = 10, 13, 16,$ and 19

distribution

$$W_\alpha^*(x) = \left(\frac{x}{\sigma}\right)^\alpha$$

with $\alpha = -1/\gamma > 0$ and $0 \leq x \leq \sigma$. Such distributions are shown in Fig. 10 with $\alpha = 5$ and $\sigma = 10$ to 19.

Having data from several inspections, we can model the distribution and the changes of the parameters α and σ. For example, we take the 50 smallest values of each inspection of the tube to estimate the parameters at each inspection. Then we consider the trend behaviour of these estimates and estimate the trend and possible behaviour of the smallest observation in a future inspection. We generated the 50 smallest data of ten inspections with changing $\sigma = 20$ (at the first inspection) which is decreasing at each inspection by 0.2 and $\alpha = 5$. The scatterplot of the 10 inspection data set is given in Fig. 11.

We estimate σ by the largest of the 50 observations of an inspection and α also by the maximum likelihood method (Table 1).

Using the simple least squares approach, we get for the trend, assuming a linear trend, the estimated line $\sigma_i = 20.0031 - 0.17845 * i$.

With this statistical approach, one can estimate the probability of a thickness value below a certain critical value. Or one can also estimate the quantile (percentile) for a very rare event, say 10^{-5} or 10^{-6}, which is related to the possible number of sites of the tube which are not sampled during the inspection. Such extreme quantiles are

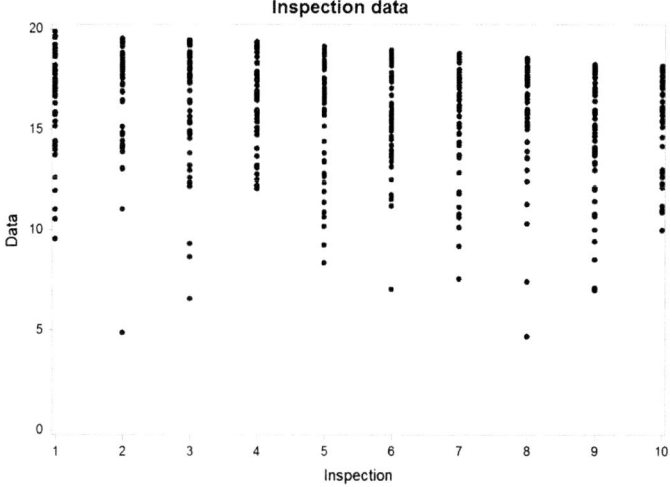

Fig. 11 Inspection data, 50 smallest observations of ten inspections with ageing effect

Table 1 Estimated α and σ from the inspection data sets

Inspection i	$\hat{\alpha}$	$\hat{\sigma}$
1	5.0663	19.885
2	5.5285	19.533
3	5.0100	19.457
4	5.4422	19.364
5	4.7389	19.136
6	5.0837	18.952
7	4.8542	18.790
8	5.4019	18.549
9	4.0119	18.248
10	6.5428	18.171

needed to discuss the safety of the whole tube or the material, and also for future inspections.

Sometimes other distributions are used for the approximation of the empirical distribution because they fit possibly slightly better, but they would have no theoretical background as the GPD distributions, which is necessary for the above extrapolation to derive extreme quantiles.

References

1. Cirillo P, Hüsler J (2009) An urn-based approach to generalized extreme shock models. Stat Probab Lett 79:969–976
2. Cirillo P, Hüsler J (2009) On the upper tail of Italian firms size distri-bution. Physica A Stat Mech Appl 388:1546–1554
3. Cirillo P, Hüsler J (2010) Shock models and firms' default: parametric and nonparametric models. In: Festschrift T. Hettmansperger, WSPC proceedings
4. Cirillo P, Hüsler J, Muliere P (2010) A nonparametric approach to interacting failing systems with an application to credit risk modeling. Intern J Theor Appl Finance 13:1223–1240
5. Cirillo P, Hüsler J (2011) Generalized extreme shock models with a possibly increasing threshold. Probab Engine Inform Sci 25:1–16
6. Cirillo P, Hüsler J (2012) An urn model for cascading failures on a lattice. Probab Engine Inform Sci 26:509–534
7. Cirillo P, Hüsler J, Muliere P (2013) Alarm systems and catastrophesfrom a diverse point of view. Method Comp Appl Probab 15:821–839
8. Cirillo P, Gallegati M, Hüsler J (2012) A Polya lattice model to study leverage dynamics and contagious financial fragility. Adv Compl Syst 15 supp. 02, 1250069 (26 pages)
9. Embrechts P, Klüppelberg C, Mikosch T (1997) Modelling extremal events for insurance and finance. Springer, Berlin
10. Falk M, Hüsler J, Reiss RD (2010) Laws of small numbers: extremesand rare events. In: DMV Seminar Band 23, 3rd edn. Birkhäuser, Basel
11. Gut A (1988) Stopped random walks. Springer, New York
12. Gut A (1990) Cumulative shock models. Adv Appl Prob 22:504–507
13. Gut A, Hüsler J (2005) Realistic variation of shock models. Stat Probab Lett 74:187–204
14. Gut A, Hüsler J (1999) Extreme shock models. Extremes 2:293–305
15. Gut A, Hüsler J (2009) Shock models. In: Nikulin MS et al (eds) Advances in Degradation Modeling: Applications to Reliability, Survival Analysis, and Finance. Series Statistics for Industry and Technology, Birkhäuser, Basel, pp 59–76
16. Gut A, Janson S (1983) The limiting behaviour of certain stopped sums and some applications. Scand J Statist 10:281–292
17. de Haan L, Ferreira A (2006) Extreme value theory. Springer Series in Operations Research and Financial Engineering. Springer, New York
18. Reiss RD, Thomas M (2007) Statistical analysis of extreme values, 3rd edn. Birkhäuser, Basel
19. Resnick S (2007) Heavy-tail phenomena. Springer, New York

Statistical Models for Interval-Censored Time-to-Event Data

Geurt Jongbloed

Abstract Time-to-event data are collected and studied in many research fields, such as medical science and reliability theory. A complication often encountered with these data is censoring. The time of the actual event is not observed: for each subject, only an interval can be observed that contains this time. Parametric, as well as nonparametric estimation, procedures can be employed to estimate the relevant quantities of interest. Also, various models can be used to include explanatory variables in the model. In this paper the parametric and nonparametric approach to interval-censored data are described. The aim is to show a glimpse of the possibilities of stochastic modelling and stimulate discussion on the development of models specifically in the context of ageing.

Keywords Maximum likelihood · EM algorithm · Accelerated life Cox proportional hazard

1 Introduction

Consider an experimental setting where a number of samples from a material are exposed to certain conditions. The question of interest is how long does it take until the quality of the material (e.g. related to strength, flexibility or other properties) drops below a certain level. Suppose the samples are 'comparable' and not obviously related to each other, say that these were obtained from one production line, but not too close together in time. Then, typically, not all samples break down at the same time. There will be variation in the observed failure times. A common approach in statistics is then to assume a stochastic model for the data. All survival times are modelled as random variables. Denote these by $X_1, X_2, \ldots, X_n$. Comparability of the objects leads to an assumption that these random variables all

G. Jongbloed (✉)
Delft Intitute of Applied Mathematics, Delft University of Technology, Delft, The Netherlands
e-mail: G.Jongbloed@tudelft.nl

© Springer International Publishing AG 2018
K. van Breugel et al. (eds.), *The Ageing of Materials and Structures*,
https://doi.org/10.1007/978-3-319-70194-3_17

have the same distribution, meaning that there exists one (distribution) function F such that for $i = 1, 2, \ldots, n$:

$$\Pr(X_i \leq x) = F(x), x \geq 0.$$

Lack of relation between the objects leads to the common assumption that the random variables are (stochastically) independent. This means that for all $i \neq j$:

$$\Pr(X_i \leq x, X_j \leq y) = \Pr(X_i \leq x) \times \Pr(X_j \leq y) = F(x) \times F(y), x, y \geq 0$$

The random variables are in this case called i.i.d., or independent and identically distributed.

Now, given a data set of measurements $\{x_1, x_2, \ldots x_n\}$, the aim is to estimate the underlying distribution function F, so to approximate this distribution function using only the available data.

In certain situations, one could impose specific assumptions on the distribution function F. For instance, that it is an exponential distribution function that can be written as:

$$F(x) = 1 - \exp(-\theta x), x \geq 0$$

for some parameter $\theta > 0$ or a Weibull distribution function that can be written as:

$$F(x) = 1 - \exp(-(\alpha x)^\beta), x \geq 0$$

for some parameter pair (α, β) with $\alpha, \beta > 0$. See Fig. 1 for a picture of some of these distribution functions (left panel) and the corresponding probability density

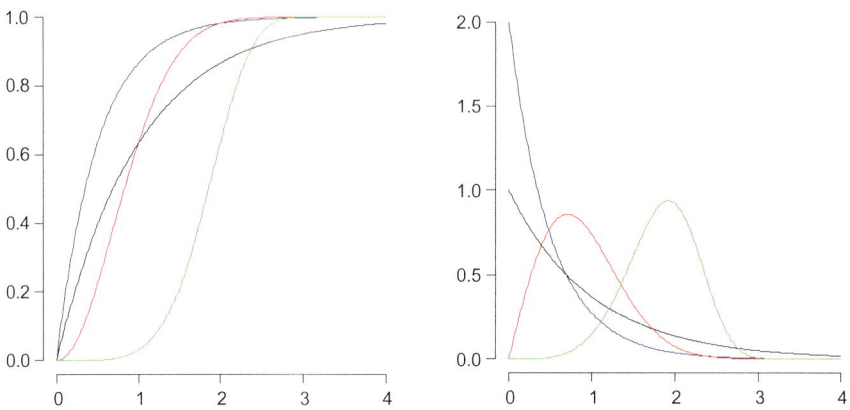

Fig. 1 The **left** panel shows four distribution functions; the **right** panel the corresponding probability density functions. Black: exponential (1), blue: exponential (2), red: Weibull (1, 2), green: Weibull (5;0,5)

functions (right panel). A probability-density function is the derivative of a distribution function.

The class of exponential (or Weibull) distribution functions is called a statistical model for the data. Under such a parametric model, estimating the distribution function boils down to statistical parameter estimation, i.e. recovering the underlying parameter $\theta > 0$ (or, in case of the Weibull model, the pair of parameters (α, β)) based on the available data.

There are general methods to construct sensible parameter estimates within a parametric model. One of these is the method of Maximum Likelihood (ML). More on this method will be treated in the next sections.

Instead of assuming a well-known parametric model, one could also use a nonparametric estimator of the distribution function. The best known estimator is undoubtedly the empirical distribution function. This distribution function (depending on the data at hand) is defined by:

$$F_n(x) = \frac{|\{i : x_i \leq x\}|}{n}$$

where $|\cdot|$ denotes the number of elements of a set. Figure 2 shows this piecewise constant empirical distribution function based on a data set of size $n = 15$. The individual points can be identified as jump points of the empirical distribution function. The figure also shows two parametric ML estimates of the distribution function: one based on the exponential model, the other on the Weibull model. If a parametric model is adequate, the ML estimator corresponding to that parametric model will be close to the empirical distribution (for reasonable sample sizes).

In Sect. 2, this general statistical approach to estimating distributions will be described for the situation where there is so-called interval censoring. In that

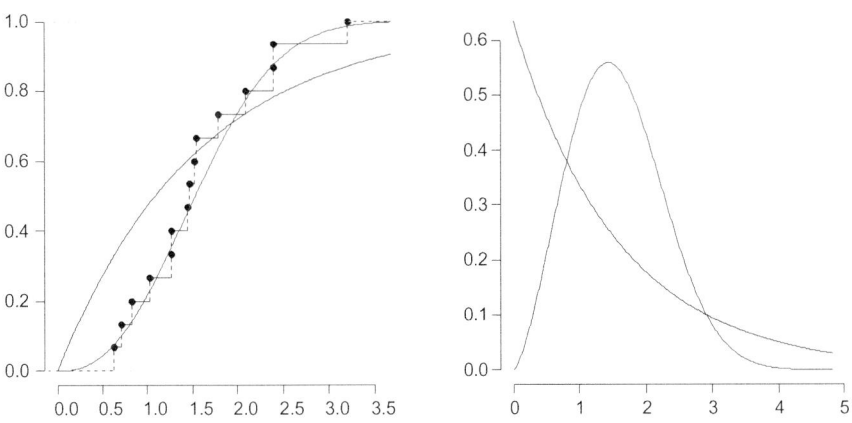

Fig. 2 Empirical distribution function of a sample of size 15, together with the parametric ML estimates: the exponential (black) and Weibull (blue). The **right** panel shows the probability-density functions corresponding to the parametric ML estimates

situation, the event times are not observed precisely: for each subject, only an interval is observed to which the corresponding event time belongs. A parametric and nonparametric approach to this problem is described. Section 3 deals with two particular types of regression model: the accelerated-life model and Cox proportional-hazard model.

2   Interval Censoring

Denote by $T_1, T_2, \ldots, T_n$ a random sample from an unknown distribution function F. Instead of observing these random variables directly, n intervals are observed,

$$(u_1, v_1], (u_2, v_2], \ldots, (u_n, v_n],$$

where the available information is that $T_i \in (u_i, v_i]$. This censoring mechanism is called interval censoring. Now, based on the observed intervals, the distribution function of interest can be estimated. Again, a model is needed, meaning a set of potential distribution functions to which the true distribution function is assumed to belong. This can be a parametric set, like the exponential distributions, but also a nonparametric set. Examples of the latter are the class of all distribution functions or the class of concave distribution functions. As in the estimation context with direct data, a natural estimation method is maximum likelihood. This approach can be taken in the parametric as well as in the nonparametric situation.

Consider a data set consisting of the intervals. For a particular distribution function F, the log likelihood can then be defined as:

$$L(F) = \sum_{i=1}^{n} \log(F(v_i) - F(u_i)).$$

The ML estimator maximizes this function over the model at hand. For the exponential model, this function reduces to:

$$\varphi(\theta) = L(F_\theta) = \sum_{i=1}^{n} \log(\exp(-\theta u_i) - \exp(-\theta v_i)).$$

For an artificial data set of size $n = 20$, visualised in the left panel of Fig. 3 the function φ is given in the right panel of this figure. Its maximiser over θ, the ML estimate, is 0,79 in this case. It is easily found by a univariate optimisation procedure.

Now, for the same data set, also a nonparametric maximum-likelihood estimator can be defined and computed. The function L is then maximised over all (sub-) distribution functions, i.e. non-decreasing functions with a range contained in $[0, 1]$. This problem is well posed and well-known algorithms like the Expectation

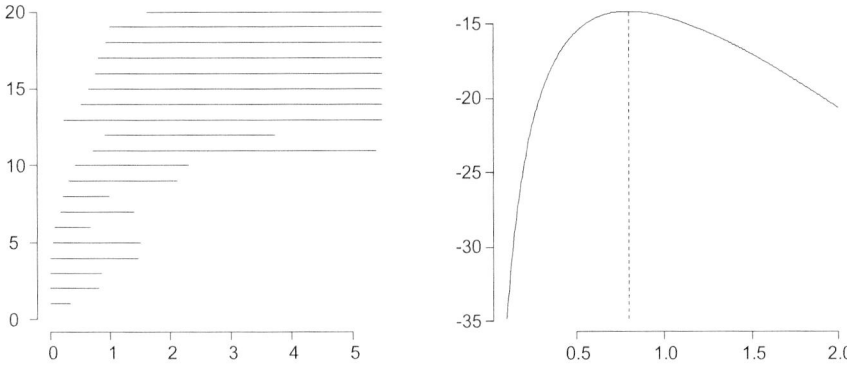

Fig. 3 Visualisation of the interval censored data. For 20 event times, intervals are shown known to contain the corresponding event time. The **right** panel shows the log likelihood function ϕ

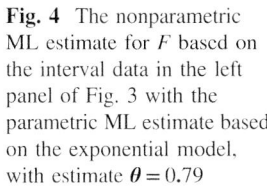

Fig. 4 The nonparametric ML estimate for F based on the interval data in the left panel of Fig. 3 with the parametric ML estimate based on the exponential model, with estimate $\theta = 0.79$

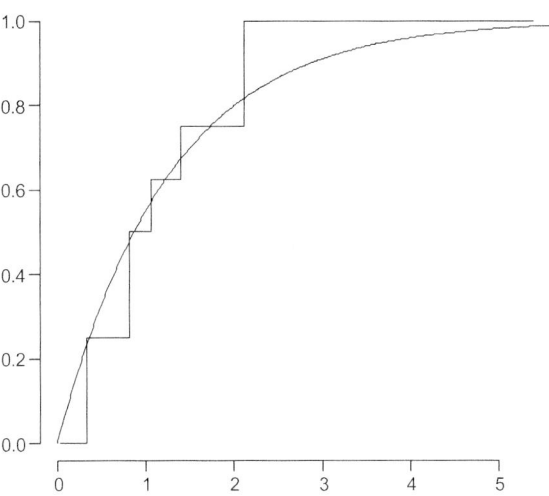

Maximization (EM) algorithm [2] and the Iterative Convex Minorant Algorithm [6]. Figure 4 shows the exponential and nonparametric ML estimator obtained based on the data of Fig. 3.

3 Regression Models with Interval Censoring

Often, instead of only observing (intervals containing) event times, additional information of interest is available for all subjects in the study. There may be some subject-specific material properties available, some measurement of thickness etc.

This extra information can be included in the model in various ways. The simplest model, including covariates, is the accelerated-life model. This model assumes the distribution functions of the n event times to be scaled versions of a single distribution function, where the scaling constant for the i-th subject is a fixed function of a linear combination of the covariate x_i. More concretely, there is a function c on $\mathbb{R}$ taking values in the positive numbers such that:

$$\Pr(T_i \leq t | x_i) = F\left(\frac{t}{c(\beta^T x_i)}\right)$$

A popular choice for the function c is $c(u) = e^u$. In the simplest situation, the covariate vector is univariate, leading to:

$$\Pr(T_i \leq t | x_i) = F\left(t e^{-\beta x_i}\right)$$

with the exponential choice for c. Using this expression for the (conditional) distribution function of the event times, the observed intervals can again be used to write down a log likelihood function:

$$L(F, \beta) = \sum_{i=1}^{n} \log(\Pr(u_i < T_i \leq v_i)) = \sum_{i=1}^{n} \log\left(F\left(\frac{v_i}{c(\beta^T x_i)}\right) - F\left(\frac{u_i}{c(\beta^T x_i)}\right)\right)$$

Another approach, to incorporate background information in the analysis, uses the hazard rate of a distribution. Given a density with distribution function F and density function f, this hazard rate (also known as failure rate) is defined by:

$$\lambda(t) = \frac{f(t)}{1 - F(t)} = -\frac{d}{dt} \log(1 - F(t)) = \lim_{\epsilon \downarrow 0} \epsilon^{-1} \Pr(T \in (t, t + \epsilon] | T > t)$$

This function quantifies the instantaneous risk of breaking down at time t, given the item has functioned until time t. Conversely, knowing the hazard rate λ, the corresponding distribution function can be obtained via:

$$F(t) = 1 - \exp\left(-\int_0^t \lambda(s) ds\right)$$

The proportional hazard model assumes the hazard rates of all items to be proportional and a specific structure of the proportionality parameters as a function of the covariate vector x:

$$\lambda_i(t) = c\left(x_i^T \beta\right) \lambda(t)$$

where $c: \mathbb{R} \to (0, \infty)$ is a known function, and λ is the so-called baseline hazard rate. The most popular model of this type is the Cox proportional-hazard model, where $c(y) = e^y$. The log likelihood of the observed intervals under this model is best written as function of β and baseline hazard λ, and given by:

$$L(\lambda, \beta) = \sum_{i=1}^{n} \log(\Pr(u_i < T_i \leq v_i))$$

$$= \sum_{i=1}^{n} \log\left(\exp\left(-c(x_i^T\beta) \int_0^{u_i} \lambda(s)ds\right) - \exp\left(-c(x_i^T\beta) \int_0^{v_i} \lambda(s)ds\right)\right)$$

The models described briefly in this section allow for great flexibility in modelling time-to-event data in the presence of interval censoring. The choice of function c, for example, could be inspired by knowledge of the material and circumstances under which the model is used. The same holds for the (possibly parametric) statistical model for the 'baseline distributions and hazards' in the various models. From a statistical point of view, estimating parameters can be quite involved, but quite a lot is possible. Recent (and forthcoming) books dealing with modelling time-to-event data, also in a regression context, include [3, 4, 7–9]. Papers that specifically deal with semiparametric regression models for interval censored event times include [5, 10].

4 Challenges with Statistics of Ageing

The aim of this paper is to show a broad audience that mathematical modelling of not precisely-observed event data is quite flexible. The simple models presented in the previous section (and more of the kind) can be extended to become more realistic. A big challenge is to come up with tailor-made extensions of the models, together with materials scientists and engineers, and implement these for practical use in specific contexts of ageing.

Experimental data on ageing of material is and can be obtained in laboratories. The circumstances can be chosen and measurements taken. Also real-life data are obtained from materials and infrastructures. Of course, there is a problem of scaling to a realistic time scale and also translating effects like 'accelerated ageing' or 'multiplied risk'. It is a challenge to design useful experiments and corresponding models that can combine real-life data obtained from a realistic time scale with experimental data and come up with lifetime predictions for infrastructures and materials.

References

1. Cox D (1972) Regression models and life-tables. J Roy Statist Soc Ser B 187–220
2. Dempster A, Laird N, Rubin D (1977) Maximum likelihood from incomplete data via the EM algorithm. J Roy Statist Soc Ser B 1–38
3. Groeneboom P, Jongbloed G (2014) Nonparametric estimation under shape constraints. Cambridge University Press
4. Hosmer DW, Lemeshow S, May S (2011) Applied survival analysis: regression modeling of time to event data. Wiley
5. Huang J (1996) Efficient estimation of the proportional hazards model with interval censoring. Ann Statist 540–568
6. Jongbloed G (1998) The iterative convex minorant algorithm for nonparametric estimation. J Comput Graph Statist 310–321
7. Kalbfleisch JD, Prentice RL (2011) The statistical analysis of failure time data. Wiley
8. Klein J, Moeschberger N (2003) Survival analysis: techniques for censored and truncated data. Springer
9. Sun J (2006) The statistical analysis of interval-censored failure time data. Springer
10. Zhang Y, Lei H, Huang J (2010) A spline-based semiparametric maximum likelihood estimation method for the Cox model with interval censored data. Scand J Statist 338–354

Erratum to:
The Ageing of Materials and Structures

Klaas van Breugel, Dessi Koleva and Ton van Beek

Erratum to:
K. van Breugel et al. (eds.), *The Ageing of Materials*
and Structures, **https://doi.org/10.1007/978-3-319-70194-3**

In the original version of the book, the incorrectly published author name "Ton Beek" has been now corrected to read as "Ton van Beek" in Cover, Copyright and Title pages.

The updated online version of the book can be found at
https://doi.org/10.1007/978-3-319-70194-3

© Springer International Publishing AG 2018   E1
K. van Breugel et al. (eds.), *The Ageing of Materials and Structures*,
https://doi.org/10.1007/978-3-319-70194-3_18

Epilogue

The AMS'14 conference on Ageing of Materials and Structures was the first of its kind event on this broad, yet, specific and challenging topic. The conference indeed proved to be a fruitful podium for lively discussions, interactions, sharing knowledge and experiences. It was a platform to address and acknowledge the worldwide dimension and significance of the subject "Ageing of Materials and Structures". As such, the event served the purpose to appreciate and judge "ageing" from various angles and viewpoints, starting from fundamental background and knowledge to practical objections and experiences, societal challenges and economic relevance.

The overall outcome was an already-improved perception on what we could call "ageing" of materials and structures, how to define it and how to control and predict it. In other words, the AMS'14 conference largely contributed to streamlining the wide range of knowledge, judgment and recognition of ageing. More importantly, the interrelationship of scientific background and reflection in practical cases, is specifically addressed, linking fundamentals to valid real-life perspectives.

Ageing—What's in the Name

During the AMS'14 conference, many definitions for "ageing of materials" were proposed. The discussions and presentations made clear that most of the authors and audience had a similar perception of ageing and spoke about it in comparable terms and manner. It appeared, however, that the currently used definition of ageing, i.e. a change of performance with elapse of time, is both debatable and controversial. Here length-scale aspects come into play. A stable performance of a material or system at the macroscale does not exclude that, at lower length scales, ageing processes have started. After all, the origin of changes in performance with time is primarily intrinsic: a material changes in relevant conditions and as a consequence from the natural alteration of material properties at lower length scales. A material will age even when it is not used at all (i.e. "materials in a state of rest").

Other factors will also influence the ageing process. The environmental conditions, together with the service or "use" of a material, will affect the ageing process

© Springer International Publishing AG 2018
K. van Breugel et al. (eds.), *The Ageing of Materials and Structures*,
https://doi.org/10.1007/978-3-319-70194-3

and can lead to enhancement of the rate of ageing. This is especially the case when a continuously changing environment (e.g. climate change, more aggressive conditions) will lead to even more significant serviceability variations of otherwise traditionally stable materials.

When reading the classic definition of ageing, the general perception is of association with the terms durability, lifetime, degradation etc. A comprehensive state-of the art exists on these subjects, encompassing research in various fields of scientific interest. Although these relevant fields of science are substantially different, the same questions often arise. These questions are mostly multidisciplinary in nature, as illustrated by chapters "Urgency and Challenges of Ageing in Science and Engineering" to "Mechanics of Ageing—From Building to Biological Materials". Therefore, they can only be answered by considering and integration of all fields of interest via a cross-border, multidisciplinary approach.

Ageing is by all means closely related to durability and service life. Durability describes how an object functions in time until the event of failure to fulfil the originally deemed requirements. The period between production of an object and the moment when it no longer conforms to the requirements by design, is called the lifetime of the object. Both the object and the initially established requirements change over time. The process of changing the material properties of an object in time is what we call ageing. These changes will occur even when the material or object still performs well! Hence, in order to thoroughly describe ageing, we need information about the mechanisms and processes that are responsible for the very onset of ageing, i.e. before the consequences of ageing emerge on the macroscale.

The changes in an object that affect negatively its performance are called 'degradation'. Ageing, however, is not only a negative effect. From an aesthetical viewpoint, ageing can also be the reason why we are attracted to certain buildings, structures or objects. Preserving these aged structures requires specialised knowledge of both preservation techniques and restoration ethics, an essential point addressed in chapter "The Relevance of Ageing for Civil Infrastructure: The Profession, the Politics, the Classroom". Accelerated ageing is a well-known production technique to fulfil some aesthetic needs for aged objects. Furthermore, in some cases, the technical performance of a material increases due to ageing. Some materials, e.g. concrete, become stronger in time and can thus bear greater loads.

A significant number of works cover purely scientific and more applied aspects of research on the durability of objects and materials. Most of this research is focussed on the changes in materials' properties over time and how these changes are reflected in their performance and serviceability. Specific research on the fundamental background of ageing-related changes is, however, rather limited. The fundamental background of ageing originates from, but also develops within, the chemical, physical, mechanical and biological changes of the material over time, together with the process of affecting material properties in concert with the progress of these changes. These interacting "components" of ageing, as a synergy or on their own, provide the essentials for a better design, as presented in chapter "Study on Triage for Deteriorated Concrete Structures by JSCE-342" of this book.

A prerequisite for an adequate and optimum design that considers ageing is a clear definition and understanding of materials' behaviour in view of risk management. Linking design and risk management in an appropriate manner is largely dependent on the mathematical models used to predict ageing and its effect on an object, structure or system. The concept of these considerations is discussed in chapter "Ageing in Shallow Underground RC Culverts and Tunnels".

The book concludes with suggestions on how to approach and pursue ageing further in view of the main objective: to save costs and minimise environmental impact by acting towards enhancement of the limited lifetime of materials and structures. In this epilogue, some conclusions are summarised as a reflection of the position of the authors and editors on recognising the complexity of ageing and the importance to pursue ageing further, targeting costs savings and minimised environmental impact.

Some Conclusions

This first conference on Aging of Materials and Structures was a truly inspiring event. A total of 237 authors within 94 contributions shared insights into how their work is related to the ageing of materials and structures. Thanks to the wide variety of contributions, from various fields of materials, applied sciences and engineering, design and simulations, the following conclusions have been drawn:

- Ageing of materials and structures is a global and widespread phenomenon. The ageing of a material affects its designed functionality, hence its performance in relevant conditions and within its service life. The risks involved due to ageing-related alterations in functionality have an implicit impact on society.
- The impact on society is mainly reflected in the potential occurrence of an undesirable event. The chance of failure can be calculated with statistics. The risk, however, is more than just chance. The impact, in terms of loss of lives, injuries, costs and discomfort, are rarely mentioned by researchers in their contributions. This despite the fact that those risks are the primary reason for the on-going research on ageing of materials and structures.
- Ageing is often approached by only describing the apparent phenomena, as observed on a material, object or structure. The solutions offered, therefore, address mostly specific ageing problems alone and can hardly be considered generic. The fundamental background of ageing is rather rarely investigated or reported in the context of the many relevant fields, such as chemistry, physics, mechanics and biology.
- When a new research theme is launched, logically, everyone is looking for the best way to accommodate the current research in the frame of the new theme and the best manner to contribute to the subject. When one reads the AMS'14 contributions, the feeling is that the authors were searching for the meaning, or the definition, of the word "ageing". Although the term itself seemed clear, the

specific meaning was open for debate. The debate was mostly about the context of the various definitions of ageing and how ageing is related to the terms durability, degradation and lifetime.

From the chapters presented at the Ageing of Materials and Structures conference in 2014 and the chapters included in this book, the drawn-up conclusion is that fundamental research is indeed required to answer the many questions raised by society and industry. Sustainability and the economics of assets are based on the lifetime of an object and the expected performance of this object during its lifetime. The materials used in an object form the backbone of these assets. How, and especially why these materials change in time, will be the basis of improvement and innovation. This requires fundamental research on the changes of materials over time, also called ageing.

The editors